A GUIDE
TO
ADVANCED REAL ANALYSIS

© 2009 by
The Mathematical Association of America (Incorporated)
Library of Congress Catalog Card Number 2009927192
ISBN 978-0-88385-343-6
Printed in the United States of America
Current Printing (last digit):
10 9 8 7 6 5 4 3 2 1

The Dolciani Mathematical Expositions
NUMBER THIRTY-SEVEN

MAA Guides # 2

A Guide

to

Advanced Real Analysis

Gerald B. Folland
University of Washington

Published and Distributed by
The Mathematical Association of America

The DOLCIANI MATHEMATICAL EXPOSITIONS series of the Mathematical Association of America was established through a generous gift to the Association from Mary P. Dolciani, Professor of Mathematics at Hunter College of the City University of New York. In making the gift, Professor Dolciani, herself an exceptionally talented and successful expositor of mathematics, had the purpose of furthering the ideal of excellence in mathematical exposition.

The Association, for its part, was delighted to accept the gracious gesture initiating the revolving fund for this series from one who has served the Association with distinction, both as a member of the Committee on Publications and as a member of the Board of Governors. It was with genuine pleasure that the Board chose to name the series in her honor.

The books in the series are selected for their lucid expository style and stimulating mathematical content. Typically, they contain an ample supply of exercises, many with accompanying solutions. They are intended to be sufficiently elementary for the undergraduate and even the mathematically inclined high-school student to understand and enjoy, but also to be interesting and sometimes challenging to the more advanced mathematician.

MAA Service Center
P.O. Box 91112
Washington, DC 20090-1112
1-800-331-1MAA   FAX: 1-301-206-9789

PREFACE

The term "real analysis" refers, in the first place, to the classical theory of functions of one and several real variables: limits and continuity, differentiation, the Riemann integral, infinite series, and related topics. However, it has come to encompass some theories of a more abstract nature that have extended the ideas of real-variable theory to much more general settings, a development which in turn has shed new light on concrete, "classical" problems. This more advanced part of real analysis is the subject of the present book.

This book is addressed, therefore, to people who are already familiar with classical real-variable theory. (Many books are available on that subject; the old classic is Rudin [16], and the most engaging of the recent ones is Körner [10]. In addition, an MAA Guide to it by Steven Krantz [11] is appearing along with this one.) In accordance with the philosophy of the MAA Guides, my aim is to give an account of the subject within a brief text that will provide an overview for the novice and a refresher for those who have already studied it. Essential definitions, major theorems, and key ideas of proofs are included; technical details are not. Thus, most of the formally stated results in the book are followed by sketches of proofs whose degree of completeness varies widely. The results for which little or no proof is provided fall into two categories, which are distinguished by the labels "Proposition" and "Theorem." If the result is called a proposition, its proof is easy, and the reader is encouraged to try it as an exercise. If it is called a theorem, its proof is long and not susceptible to condensation into a short sketch.

Of course, this presentation works only if the reader has a resource for filling in the gaps. I take my own book [6] as a standard reference for a more complete account of the material in this book, simply because I am most familiar with it. All the results stated here are proved in [6] except those for which an explicit reference is given to some other source. Lang

[12], Royden [15], and Rudin [17] are other books that cover most of the same material.

This book is not, however, merely a condensed version of [6]. One of the main problems for a textbook writer, as for a novelist or historian, is to figure out a way of turning a body of material whose parts have many interconnections into a linear narrative, and the solution is generally far from unique. I have taken the opportunity afforded by the nature of MAA Guides to arrange the topics in a quite different way than I did in [6]. Perhaps readers who examine both texts will gain something from comparing the two perspectives.

Gerald B. Folland
Seattle, April 2009

Contents

PROLOGUE

NOTATION, TERMINOLOGY, AND SET THEORY

In this prologue we set the stage by briefly discussing some points of notation and terminology and a few facts from set theory that will be used throughout the book.

NUMBERS

We set

$$\mathbb{N} = \text{the set of positive integers,}$$
$$\mathbb{Z} = \text{the set of integers,}$$
$$\mathbb{R} = \text{the set of real numbers,}$$
$$\mathbb{C} = \text{the set of complex numbers.}$$

We often enlarge the real number system by adjoining two "elements at infinity," ∞ (also called $+\infty$ for emphasis) and $-\infty$. In the extended system $\mathbb{R} \cup \{\pm\infty\} = [-\infty, \infty]$, every set E has a least upper bound or *supremum* and a greatest lower bound or *infimum*, denoted respectively by $\sup E$ and $\inf E$. Moreover, every infinite series with nonnegative terms has a well-defined sum in $[0, \infty]$, namely, the supremum of its partial sums.

If $z = x + iy$ is a complex number, its complex conjugate $x - iy$ is denoted by $\bar{z}$, and its absolute value or modulus $\sqrt{z\bar{z}} = \sqrt{x^2 + y^2}$ is denoted by $|z|$.

The spaces of ordered n-tuples of real or complex numbers are denoted by $\mathbb{R}^n$ and $\mathbb{C}^n$. If $u = (u_1, \ldots, u_n)$ belongs to $\mathbb{R}^n$ or $\mathbb{C}^n$, we denote its Euclidean norm by $|u|$:

$$|u| = \left[\sum_1^n |u_j|^2 \right]^{1/2}.$$

We also define the dot product of two elements u, v of $\mathbb{R}^n$ by

$$u \cdot v = \sum_1^n u_j v_j.$$

SETS AND MAPPINGS

We employ standard notation from set theory. The set inclusion sign $\subset$ is interpreted in the wide sense; that is, the condition $E \subset F$ allows the possibility that $E = F$. We denote the relative complement of F in E by $E \setminus F$:

$$E \setminus F = \{x \in E : x \notin F\}.$$

We denote the empty set by $\varnothing$. A family $\{E_\alpha\}_{\alpha \in A}$ of sets is called *disjoint* if $E_\alpha \cap E_\beta = \varnothing$ whenever $\alpha \neq \beta$.

When it is understood that we are considering subsets of a fixed set X, we may speak simply of the complement of a set E (in X):

$$E^c = X \setminus E.$$

In this situation we have *De Morgan's laws*: If $\{E_\alpha\}_{\alpha \in A}$ is a collection of subsets of X, then

$$\left[\bigcup_{\alpha \in A} E_\alpha\right]^c = \bigcap_{\alpha \in A} E_\alpha^c, \qquad \left[\bigcap_{\alpha \in A} E_\alpha\right]^c = \bigcup_{\alpha \in A} E_\alpha^c.$$

We denote the collection of all subsets of X (including X and $\varnothing$) by $\mathcal{P}(X)$.

Suppose X and Y are nonempty sets. In strict set-theoretic terms, a *map* or *mapping* from X to Y is a collection f of ordered pairs (x, y) with $x \in X$ and $y \in Y$, such that for each $x \in X$ there is a unique $y \in Y$ (denoted by $f(x)$) with $(x, y) \in f$. (Of course, in more informal terms, we usually think of a map as a "rule" that assigns to each $x \in X$ an element $f(x)$ of Y.) A map $f : X \to Y$ is called *injective* if $f(x_1) = f(x_2)$ only when $x_1 = x_2$, *surjective* if $\{f(x) : x \in X\} = Y$, and *bijective* if it is both injective and surjective. When we wish to describe a map without giving it a name, we use the notation $x \mapsto y$ to indicate that y is the image of x under the map; for example, the squaring function on $\mathbb{R}$ is $x \mapsto x^2$.

Each map $f : X \to Y$ induces a map, still denoted by f, from $\mathcal{P}(X)$ to $\mathcal{P}(Y)$,

$$f(E) = \{f(x) : x \in E\},$$

as well as a map denoted by f^{-1} from $\mathcal{P}(Y)$ to $\mathcal{P}(X)$:

$$f^{-1}(E) = \{x : f(x) \in E\}.$$

It is an important fact that the inverse-image map $f^{-1} : \mathcal{P}(Y) \to \mathcal{P}(X)$ preserves unions, intersections, and complements:

$$f^{-1}\left[\bigcup_{\alpha \in A} E_\alpha\right] = \bigcup_{\alpha \in A} f^{-1}(E_\alpha), \qquad f^{-1}\left[\bigcap_{\alpha \in A} E_\alpha\right] = \bigcap_{\alpha \alpha \in A} f^{-1}(E_\alpha),$$
$$f^{-1}(E^c) = [f^{-1}(E)]^c.$$

(The direct-image map $f : \mathcal{P}(X) \to \mathcal{P}(Y)$ preserves unions, but it fails to preserve intersections when f is not injective, and it fails to preserve complements when f is not bijective.)

Let $\{X_\alpha\}_{\alpha \in A}$ be an indexed collection of sets. The *Cartesian product* of the sets X_α, denoted by $\prod_{\alpha \in A} X_\alpha$, is the set of all maps f from A into $\bigcup_{\alpha \in A} X_\alpha$ such that $f(\alpha) \in X_\alpha$ for all α:

$$\prod_{\alpha \in A} X_\alpha = \left\{ f : A \to \bigcup_{\alpha \in A} X_\alpha : f(\alpha) \in X_\alpha \text{ for all } \alpha \in A \right\}.$$

If $X = \prod_{\alpha \in A} X_\alpha$ and $\alpha \in A$, the αth *coordinate map* $\pi_\alpha : X \to X_\alpha$ is defined by $\pi_\alpha(f) = f(\alpha)$; we often write x and x_α instead of f and $f(\alpha)$ and call x_α the αth coordinate of x.

ZORN'S LEMMA

Every so often, especially when one is working in a very general context, one needs a theorem asserting the existence of some mathematical object but has no way of producing it by explicit construction. Often the stratagem needed to resolve the question is one of a group of related principles of general set theory pertaining to partially ordered sets. Here are the necessary definitions.

A *partially ordered set* is a set X equipped with a binary relation $\leq$ with the following properties:

i. If $x \leq y$ and $y \leq z$ then $x \leq z$.

ii. If $x \leq y$ and $y \leq x$, then $x = y$.

iii. $x \leq x$ for all x.

A partially ordered set X is called *linearly ordered* if it also satisfies

iv. If $x, y \in X$, then either $x \leq y$ or $y \leq x$.

A *maximal element* of a partially ordered set is an element x such that the only element y with $x \leq y$ is x itself.

For example, $\mathbb{R}$ is linearly ordered by the usual ordering $\leq$, and for any set S, $\mathcal{P}(S)$ is partially ordered by the inclusion relation $\subset$. If X is the collection of all proper subsets of S, partially ordered by inclusion, its maximal elements are the subsets whose complement consists of a single point. The collection of all finite subsets of an infinite set has no maximal elements.

The general existence principle most often invoked is known as *Zorn's lemma*:

Zorn's lemma. *If X is a partially ordered set and every linearly ordered subset L of X has an upper bound (i.e., an element $x \in X$ such that $y \leq x$ for all $y \in L$), then X has a maximal element.*

An alternative formulation, known as the *Hausdorff maximal principle*, is that *every partially ordered set has a maximal linearly ordered subset.* (Indeed, an upper bound for a maximal linearly ordered subset of X is a maximal element of X. On the other hand, an application of Zorn's lemma to the collection of linearly ordered subsets of X, which is partially ordered by inclusion, yields a maximal linearly ordered subset.)

Another general existence principle is the *axiom of choice*, which says that if $\{X_\alpha\}_{\alpha \in A}$ is a nonempty collection of nonempty sets, one can form a new set Y by picking one element from each X_α. Since the range of any element of the Cartesian product $\prod_{\alpha \in A} X_\alpha$ is such a set, one can state the axiom of choice as follows:

The axiom of choice. *The Cartesian product of any nonempty collection of nonempty sets is nonempty.*

Zorn's lemma implies the axiom of choice. (Consider the collection $\mathcal{F}$ of all mappings f from subsets of A into $\bigcup_{\alpha \in A} X_\alpha$ such that $f(\alpha) \in X_\alpha$ for all α in the domain of f, which is partially ordered by extension: $f \leq g$ if $\mathrm{dom}(g) \supset \mathrm{dom}(f)$ and $g|\mathrm{dom}(f) = f$.) One can also prove Zorn's lemma from the axiom of choice, but the argument is more involved. (For this and related matters, a good reference is Halmos [7].) Neither of these principles can be deduced from the other standard axioms of set theory.

The use of nonconstructive existence principles such as Zorn's lemma has not been without controversy. However, most mathematicians take the attitude that they are perfectly legitimate, while recognizing that constructive methods tend to be more informative when they are available.

CHAPTER 1

TOPOLOGY

The subject of this chapter is *point-set topology* or *general topology*, the abstract mathematical framework for the study of limits, continuity, and the related geometric properties of sets.

1.1 METRIC SPACES

The early years of the twentieth century witnessed a great increase in the level of abstraction and generality in mathematical thinking. In particular, mathematicians at that time developed theories that provide a very general setting for studying the circle of ideas related to limits and continuity, which previously had been considered in the context of subsets of Euclidean space or functions of one or several real or complex variables.

The most straightforward generalization of Euclidean space for this purpose is the notion of a metric space. A *metric space* is a nonempty set X equipped with a function

$$\rho : X \times X \rightarrow [0, \infty)$$

(a *metric*) that satisfies the following three conditions:
  i. $\rho(x, y) = \rho(y, x)$ for all $x, y \in X$.
  ii. (The *triangle inequality*) $\rho(x, z) \leq \rho(x, y) + \rho(y, z)$ for all $x, y, z \in X$.
  iii. $\rho(x, y) = 0$ if and only if $x = y$.
One should think of $\rho(x, y)$ as being the distance from x to y. We speak of "the metric space (X, ρ)," or, if ρ is understood, just "the metric space X."

Some examples of metric spaces:

- X = a subset of $\mathbb{R}^n$; $\rho(x, y) = \left[\sum_1^n |x_j - y_j|^2 \right]^{1/2}$ (the Euclidean distance from x to y).

5

- X = the unit sphere in $\mathbb{R}^3$; $\rho(x, y)$ = the great circle distance from x to y.

- X = a smooth curve in $\mathbb{R}^n$; $\rho(x, y)$ = the length of the arc from x to y. (That is, X is the image of a one-to-one map $\phi : (a, b) \to \mathbb{R}^n$ of class $C^{(1)}$, and if $x = \phi(s)$ and $y = \phi(t)$, $\rho(x, y) = \int_s^t |\phi'(u)|\, du$.)

- X = the set of continuous real-valued functions on the interval $[0, 1]$; $\rho(f, g) = \sup_{t\in[0,1]} |f(t) - g(t)|$.

- X = the set of continuous real-valued functions on the interval $[0, 1]$; $\rho(f, g) = \int_0^1 |f(t) - g(t)|\, dt$.

- X = any nonempty set; $\rho(x, y) = 1$ for all $x, y \in X$ with $x \neq y$.

- $X = X_1 \times X_2$ where (X_1, ρ_1) and (X_2, ρ_2) are metric spaces;

$$\rho\big((x_1, x_2),\ (y_1, y_2)\big) = \max\big(\rho_1(x_1, y_1),\ \rho_2(x_2, y_2)\big).$$

In a metric space (X, ρ), the notion of distance between two points generalizes to give a notion of distance between a point x and a set E, or between two sets E and F, as follows:

$$\rho(x, E) = \inf\{\rho(x, y) : y \in E\},$$
$$\rho(E, F) = \inf\{\rho(x, y) : x \in E,\ y \in F\}.$$

Much of the standard terminology for subsets of Euclidean spaces generalizes directly to metric spaces. In the following list of definitions, (X, ρ) is a metric space and E is a subset of X.

- If $x \in X$ and $r > 0$, the (open) *ball of radius r about x* is

$$B(r, x) = \{y \in X : \rho(x, y) < r\}.$$

- If $x \in E$, x is an *interior point* of E, or E is a *neighborhood* of x, if $B(r, x) \subset E$ for some $r > 0$.

- A point $x \in X$ is an *accumulation point* (or *limit point*) of E if the punctured ball $B(r, x) \setminus \{x\}$ contains points of E for every $r > 0$.

- E is *open* if all of its points are interior points. (This condition is vacuously satisfied if E has no points, i.e., if $E = \varnothing$.)

- E is *closed* if it contains all of its accumulation points. (Again, this condition is vacuously satisfied if E has no accumulation points.)

Concerning open and closed sets, we have the following facts, which are easy consequences of the definitions of open and closed sets, the triangle inequality, and De Morgan's laws.

1.1 Proposition. *Let (X, ρ) be a metric space.*
a. *$B(r, x)$ is an open set for every $x \in X$ and $r > 0$.*
b. *$E \subset X$ is open if and only if it is a union of balls.*
c. *The union of any family of open sets is open.*
d. *The intersection of any finite family of open sets is open.*
e. *$E \subset X$ is closed if and only if its complement $E^c = X \setminus E$ is open.*
f. *The intersection of any family of closed sets is closed.*
g. *The union of any finite family of closed sets is closed.*

Here is some more terminology concerning subsets E of a metric space (X, ρ):

- The *interior* of E is the largest open subset of E: namely, the union of all the balls contained in E.

- The *closure* of E is the smallest closed set containing E: namely, the complement of the interior of E^c, or the union of E and its set of accumulation points; it is denoted by $\overline{E}$.

- The set E is *dense* in X if $\overline{E} = X$.

- The set E is *nowhere dense* if the interior of $\overline{E}$ is empty.

- The space X is *separable* if it has a countable dense subset.

The notions of continuous mappings and sequential convergence generalize easily to mappings between metric spaces (X, ρ) and (Y, σ):

- A mapping $f : X \to Y$ is *continuous* at $x_0 \in X$ if for every $\epsilon > 0$ there is a $\delta > 0$ such that $\sigma(f(x), f(x_0)) < \epsilon$ whenever $\rho(x, x_0) < \delta$.

- A mapping $f : X \to Y$ is *continuous* if it is continuous at every point of X.

- A sequence $\{x_n\}$ in X *converges to* $x \in X$ if for every $\epsilon > 0$ there is an integer N such that $\rho(x_n, x) < \epsilon$ for all $n > N$.

- A sequence $\{x_n\}$ in X is called *Cauchy* if for every $\epsilon > 0$ there is an integer N such that $\rho(x_m, x_n) < \epsilon$ for all $m, n > N$.

- X is *complete* if every Cauchy sequence in X converges.

Much of the elementary analysis of limits, continuity, and the geometry of sets can be carried over from Euclidean spaces to general metric spaces with little change. Here are three basic results of this sort:

1.2 Proposition. *A map* $f : X \to Y$ *between two metric spaces is continuous at* $x_0 \in X$ *if and only if* $f(x_n) \to f(x_0)$ *whenever* $x_n \to x_0$.

1.3 Proposition. *A map* $f : X \to Y$ *is continuous if and only if* $f^{-1}(U)$ *is open in* X *for every open* $U \subset Y$.

1.4 Proposition. *A point* x *in a metric space* X *belongs to the closure of a set* $E \subset X$ *if and only if there is a sequence in* E *that converges to* x, *and it is an accumulation point of* E *if and only if there is a sequence in* $E \setminus \{x\}$ *that converges to* x.

For much of this analysis, however, a metric provides more precise information than is really needed. In particular, one can almost always replace a metric ρ by any other metric ρ' that is *equivalent* to it in the sense that for some constants $C_1, C_2 > 0$ we have

$$C_1 \rho(x, y) \leq \rho'(x, y) \leq C_2 \rho(x, y) \quad \text{for all } x, y \in X.$$

For example, on $\mathbb{R}^n$ it is sometimes convenient to replace the Euclidean metric ρ_E by $\rho_1(x, y) = \sum |x_j - y_j|$ or $\rho_2(x, y) = \max_j |x_j - y_j|$; these are all equivalent since

$$\rho_2 \leq \rho_E \leq \rho_1 \leq \sqrt{n}\, \rho_E \leq n \rho_2.$$

For many purposes even more distortion is permissible: the metric

$$(1.5) \qquad\qquad\qquad \rho(x, y) = |e^x - e^y|$$

on $\mathbb{R}$ is not equivalent to the standard metric $|x - y|$, but it defines the same open and closed sets, the same convergent sequences, and the same continuous mappings (with $\mathbb{R}$ as either domain or target space), simply because the exponential function and its inverse are both continuous.

In fact, what is essential for discussing limits and continuity is not the metric but the open sets that it defines, for the ϵ-δ inequalities in the definition of continuity or sequential convergence can be rephrased in terms of neighborhoods of points. Moreover, there are situations, most often involving sets of functions rather than subsets of geometric objects, where one can define useful notions of neighborhoods and open sets that do not arise from a metric. These observations lead to the more general notion of a topological space, which we shall study in the remainder of this chapter.

The one essential part of the theory of metric spaces that really requires the use of a particular metric (or, rather, equivalence class of metrics) is that relating to Cauchy sequences and completeness. For example, $\mathbb{R}$ is complete with the usual metric but not with the metric defined by (1.5): the sequence $\{-n\}_1^\infty$ is Cauchy with respect to the latter metric, but it has no limit.

1.2 TOPOLOGICAL SPACES AND CONTINUOUS MAPS

Following up on the ideas at the end of the preceding section, and using Proposition 1.1(c,d) as a guide, we construct an abstract framework for discussing open sets. Suppose X is a nonempty set. A *topology* on X is a collection $\mathcal{T}$ of subsets of X with the following properties:

  i. $\varnothing \in \mathcal{T}$ and $X \in \mathcal{T}$.
 ii. If $\{U_\alpha\}_{\alpha \in A}$ is an arbitrary collection of sets in $\mathcal{T}$, then $\bigcup_{\alpha \in A} U_\alpha \in \mathcal{T}$.
iii. If $\{U_j\}_{j=1}^n$ is a finite collection of sets in $\mathcal{T}$, then $\bigcap_{j=1}^n U_j \in \mathcal{T}$.

A set X equipped with a topology $\mathcal{T}$ is called a *topological space*; we often speak of "the topological space X" when $\mathcal{T}$ is understood.

Sometimes one needs to consider two or more different topologies on a set X. If $\mathcal{T}_1$ and $\mathcal{T}_2$ are topologies on X, we say that $\mathcal{T}_1$ is *weaker* than $\mathcal{T}_2$, or that $\mathcal{T}_2$ is *stronger* than $\mathcal{T}_1$, if $\mathcal{T}_1 \subset \mathcal{T}_2$.

Here are some classes of examples:

- If X is any metric space, the collection of all open sets in X is a topology on X.

- If X is any nonempty set, the collection of all subsets of X is a topology on X, called the *discrete topology*. (It is the topology associated to the metric defined by $\rho(x, y) = 1$ for all $x \neq y$.)

- Suppose X is a nonempty set and $\mathcal{E}$ is any collection of subsets of X. In view of the preceding example, there is at least one topology on X that contains all the members of $\mathcal{E}$. The intersection of all such topologies is again a topology; it is the weakest topology that contains all the members of $\mathcal{E}$, and it is called the topology *generated by* $\mathcal{E}$. It is not hard to show that this topology consists of $\varnothing$, X, and the collection of all unions of finite intersections of members of $\mathcal{E}$. For example, the topology associated to a metric on X is generated by the set of balls $B(r, x)$ ($r > 0$, $x \in X$). We shall exhibit some other useful examples of this construction later in this section.

- If $(X, \mathfrak{T})$ is a topological space and Y is a subset of X, $\{U \cap Y : U \in \mathfrak{T}\}$ is a topology on Y, called the *relative topology*.

If $(X, \mathfrak{T})$ is a topological space, the members of $\mathfrak{T}$ are called *open sets*. Most of the terminology and results from the preceding section can be adapted to topological spaces if one makes suitable adjustments so that the open sets, rather than the metric, are the primitive data. In particular:

- If $x \in X$ and $E \subset X$, x is an *interior point* of E, or E is a *neighborhood* of x, if there is an open set U such that $x \in U \subset E$. (In particular, if E is itself open, we can take $U = E$, so every point of E is an interior point.)

- If $x \in X$ and $E \subset X$, x is an *accumulation point* of E if, for every open set U such that $x \in U$, the set $U \setminus \{x\}$ contains points of E.

- A set $E \subset X$ is *closed* if its complement $X \setminus E$ is open, or equivalently if E contains all of its accumulation points.

- The notions of the interior and closure of a set, dense and nowhere dense sets, and separable spaces are defined just as in §1.1.

Proposition 1.1(f,g) is valid in general topological spaces: the intersection of any family of closed sets is closed, and the union of a finite family of closed sets is closed. This is immediate from the defining conditions for a topology and the characterization of closed sets as the complements of open sets.

The notions of continuity and sequential convergence (but not Cauchy sequences or completeness) also generalize readily:

- If X and Y are topological spaces, a map $f : X \to Y$ is *continuous at* $x \in X$ if $f^{-1}(U)$ is a neighborhood of x whenever U is a neighborhood of $f(x)$; f is *continuous* (with no qualification) if it is continuous at every point.

- A sequence $\{x_n\}$ in a topological space X *converges* to $x \in X$ if for every neighborhood U of x there is an integer N such that $x_n \in U$ for $n > N$.

Proposition 1.3 remains valid in this setting: A map $f : X \to Y$ is continuous if and only if $f^{-1}(U)$ is open in X whenever U is open in Y. This characterization of continuity shows that it belongs to the class of structure-preserving maps between various kinds of mathematical objects, such as linear maps between vector spaces or homomorphisms between groups or rings. In the latter situations, the structure in question consists of algebraic operations such as addition and multiplication, and the structure-preserving maps take the operations on the domain space to the corresponding operations on the target space. But here the structure is the collection of open sets of the space in question, and it is f^{-1} rather than f itself that commutes with the Boolean operations of union, intersection, and complement, so f^{-1} is the natural carrier of the structure from one space to another. In the language of category theory, the continuous maps are the *morphisms* in the category of topological spaces.

The corresponding notion of *isomorphism* here is that of a bijective map $f : X \to Y$ such that f and f^{-1} are both continuous, so that f induces a bijection between the open sets in Y and the open sets in X. Such maps are called *homeomorphisms*, and two spaces X and Y that admit a homeomorphism are called *homeomorphic*.

In many situations it is natural to consider topologies that are defined expressly in order to ensure the continuity of certain maps. Suppose X is a nonempty set, $\{Y_\alpha\}_{\alpha \in A}$ is a family of topological spaces, and $\{\phi_\alpha : X \to Y_\alpha\}_{\alpha \in A}$ is a family of maps. The topology generated by the sets $\phi_\alpha^{-1}(U_\alpha)$, as U_α ranges over the open sets in Y_α and α ranges over A, is the weakest topology on X that makes all the maps ϕ_α continuous; it is called the *topology generated by* the family $\{\phi_\alpha\}_{\alpha \in A}$.

For example, let S be any nonempty set, and let X be the set of all real-valued functions on S. For each $s \in S$ we have the evaluation map $\phi_s : X \to \mathbb{R}$ defined by $\phi_s(f) = f(s)$. The topology generated by this family of maps is called the *topology of pointwise convergence*, a terminology justified by the following result.

1.6 Proposition. *A sequence $\{f_n\}$ of real-valued functions on S converges to a function f with respect to the topology of pointwise convergence if and only if $\lim_{n \to \infty} f_n(s) = f(s)$ for every $s \in S$.*

When S is uncountable, the topology of pointwise convergence does not come from a metric: this is a classic example of the utility of working in the more general category of topological spaces.

A more general construction is the so-called product topology on the Cartesian product of a family of topological spaces. Let $\{X_\alpha\}_{\alpha \in A}$ be a family of topological spaces, and let $X = \prod_{\alpha \in A} X_\alpha$. (Recall that X is the set of all mappings $x : A \to \bigcup_{\alpha \in A} X_\alpha$ such that $x(\alpha) \in X_\alpha$ for all α.) For each $\alpha \in A$ we have the canonical projection or coordinate map $\pi_\alpha : X \to X_\alpha$ defined by $\pi_\alpha(x) = x(\alpha)$; the *product topology* on X is the topology generated by these coordinate maps. It consists of all unions of sets of the form

$\bigcap_1^n \pi_{\alpha_j}^{-1}(U_j)$ where n is an arbitrary positive integer and U_j is an open set in X_{α_j} for $j = 1, \ldots, n$. (When the index set A is finite, the product topology is generated by sets of the form $\prod_{\alpha \in A} U_\alpha$ where U_α is open in X_α, because $\prod_{\alpha \in A} U_\alpha = \bigcap_{\alpha \in A} \pi_\alpha^{-1}(U_\alpha)$. One can consider the topology generated by such products of open sets even when A is infinite, but it turns out to be less useful than the product topology.)

The notion of a topology on a set X is extremely general, and it includes examples of little interest such as the topology whose only elements are $\varnothing$ and X. For most purposes it is desirable to impose additional conditions that guarantee that there are "sufficiently many" open sets. The two most important conditions of this kind are as follows.

- A *Hausdorff space* is a topological space X such that for any two distinct points x and y in X there are disjoint open sets U and V such that $x \in U$ and $y \in V$.

- A *normal space* is a Hausdorff space X such that for any two disjoint closed sets E and F in X there are disjoint open sets U and V such that $E \subset U$ and $F \subset V$.

The Hausdorff condition probably seems so natural that one might wonder why it is worthwhile to consider more general topological spaces. In fact, there are situations, in subjects as diverse as algebraic geometry and harmonic analysis, where non-Hausdorff topologies arise naturally, but we shall say no more about them here.

Every metric space is normal. Indeed, if E and F are disjoint closed sets, it is easily verified that

$$U = \{x : \rho(x, E) < \rho(x, F)\} \quad \text{and} \quad V = \{x : \rho(x, F) < \rho(x, E)\}$$

are disjoint open sets containing E and F, respectively.

The most important feature of normal spaces is that they have a rich supply of continuous functions. More precisely, we have the following fundamental result.

1.7 Theorem. *If X is a topological space, the following conditions are equivalent.*

a. *X is normal.*

b. *For any two disjoint closed sets A and B in X there is a continuous $f : X \to [0, 1]$ such that $f(x) = 1$ for $x \in A$ and $f(x) = 0$ for $x \in B$.*

c. *For any closed set E in X and any $\phi : E \to \mathbb{R}$ that is continuous with respect to the relative topology on E, there is a continuous $f : X \to \mathbb{R}$ such that $f|E = \phi$. Moreover, if $\phi(E) \subset [a, b]$, f can be chosen so that $f(X) \subset [a, b]$.*

The implication (a) $\Rightarrow$ (b) is known as **Urysohn's lemma**, and the implication (a) $\Rightarrow$ (c) is known as the **Tietze extension theorem**, although both of them are due to Urysohn. (Tietze did only the case $X = \mathbb{R}^2$.) Urysohn proceeded by showing that (a) $\Rightarrow$ (b) $\Rightarrow$ (c); up to minor variations, his argument is still the standard one. The reverse implications are easy: (b) is the special case of (c) where $E = A \cup B$ and $f : E \to [0, 1]$ is defined by $f = 1$ on A, $f = 0$ on B; and if f is as in (b), the sets $f^{-1}((\frac{1}{2}, \infty))$ and $f^{-1}((-\infty, \frac{1}{2}))$ are disjoint open sets containing A and B respectively, so (a) follows.

1.3 NEIGHBORHOOD BASES AND CONVERGENCE

The notion of sequential convergence plays a central role in analysis on metric spaces, but its utility is somewhat more limited in more general topological spaces. In particular, Proposition 1.4 is only half true in the more general setting: if E is a set in a topological space X and $x \in X$, it is true that if there is a sequence $\{x_n\}$ in E that converges to x then x is in the closure of E, but the converse is false. For example, let X be the set of all real-valued functions on $[0, 1]$, with the topology of pointwise convergence, and let $E \subset X$ be the set of all continuous functions on $[0, 1]$. Then E is dense in X: for any $f \in X$ and any finite set $\{t_1, \ldots, t_n\} \subset [0, 1]$ there are continuous functions g such that $g(t_j) = f(t_j)$ for all j, so every neighborhood of f contains continuous functions. However, not every function on $[0, 1]$ is the pointwise limit of a sequence of continuous functions. One way to see this is to invoke the fact that the limit of a pointwise convergent sequence of continuous functions must be Borel measurable (see §2.2). Another way is to use a cardinality argument: the set of sequences of continuous functions has the cardinality c of the continuum, but the set of all functions has cardinality 2^c, which is strictly larger.

The reason why sequential convergence does not perform well in this example is that the set of neighborhoods of a function on $[0, 1]$ (or on any uncountable set) in the topology of pointwise convergence is much more complicated than the set of balls centered at a point in a metric space. Some definitions are in order pertaining to a topological space $(X, \mathcal{T})$:

- A *neighborhood base* for $\mathcal{T}$ at $x \in X$ is a family $\mathcal{N} \subset \mathcal{T}$ such that (i) $x \in U$ for all $U \in \mathcal{N}$, and (ii) for every neighborhood V of x there is a $U \in \mathcal{N}$ such that $U \subset V$.

- A *base* for $\mathcal{T}$ is a family $\mathcal{B} \subset \mathcal{T}$ that contains a neighborhood base at each point of X.

- $(X, \mathcal{T})$ is called *first countable* if there is a countable neighborhood base for $\mathcal{T}$ at each $x \in X$.

- $(X, \mathcal{T})$ is called *second countable* if there is a countable base for $\mathcal{T}$.

Some examples:

- If $\mathcal{T}$ is generated by a family of sets $\mathcal{E}$, the set of intersections of finite subfamilies of $\mathcal{E} \cup \{X\}$ is a base for $\mathcal{T}$. In particular, if $\mathcal{E}$ is countable, then so is this base, so $(X, \mathcal{T})$ is second countable.

- Every metric space is first countable, for the balls of rational radius centered at x are a countable neighborhood base for the topology at x.

- Every separable metric space is second countable, for the balls of rational radius centered at the points in a countable dense set are a countable base for the topology.

- Conversely, every second countable topological space is separable. (If $\mathcal{B}$ is a countable base for the topology, pick a point x_B in each nonempty $B \in \mathcal{B}$; then $\{x_B : B \in \mathcal{B}\}$ is a countable dense set.)

First countability is precisely what one needs in order to make sequential convergence work in the familiar way. In particular, Proposition 1.4 remains valid for all first countable spaces. For spaces that are not first countable, the resolution of the difficulty is to generalize the notion of sequence by allowing index sets more general than the positive integers that can be adapted to the particular problem at hand. Here are the definitions.

- A *directed set* is a set A equipped with a binary relation $\lesssim$ that is reflexive ($\alpha \lesssim \alpha$ for all $\alpha \in A$) and transitive (if $\alpha \lesssim \beta$ and $\beta \lesssim \gamma$ then $\alpha \lesssim \gamma$), such that for every $\alpha, \beta \in A$ there is a $\gamma \in A$ such that $\alpha \lesssim \gamma$ and $\beta \lesssim \gamma$.

- A *net* in a set X indexed by a directed set A is a mapping $\alpha \mapsto x_\alpha$ from A into X. We denote nets indexed by A by $\langle x_\alpha \rangle_{\alpha \in A}$.

- If X is a topological space, a net $\langle x_\alpha \rangle_{\alpha \in A}$ in X *converges* to $x \in X$ (in symbols: $x_\alpha \to x$) if for every neighborhood U of x there is an $\alpha_0 \in A$ such that $x_\alpha \in U$ whenever $\alpha_0 \lesssim \alpha$.

- If X is a topological space, a point $x \in X$ is a *cluster point* of the net $\langle x_\alpha \rangle_{\alpha \in A}$ if for every neighborhood U of x and every $\alpha \in A$ there is a $\beta \in A$ such that $\alpha \lesssim \beta$ and $x_\beta \in U$.

The simplest example of a directed set is the set $\mathbb{N}$ of positive integers with its usual ordering, and a net indexed by $\mathbb{N}$ is simply a sequence. Another example of a directed set that is familiar from the theory of the Riemann integral is the set of partitions P of an interval $[a, b]$ into finitely many subintervals, with $P \lesssim Q$ if and only if the maximum length of the subintervals in P is no less than the maximum length of the subintervals in Q. The fundamental example that lies at the heart of the theory of nets in a topological space X is the set of all neighborhoods of a point $x \in X$, with $U \lesssim V$ if and only if $U \supset V$. With this in mind, it is not hard to establish the following generalizations of Propositions 1.2 and 1.4.

1.8 Proposition. *A map $f : X \to Y$ between two topological spaces is continuous at $x_0 \in X$ if and only if $f(x_\alpha) \to f(x_0)$ for every net $\langle x_\alpha \rangle$ in X that converges to x_0.*

1.9 Proposition. *A point x in a topological space X belongs to the closure of a set $E \subset X$ if and only if there is a net in E that converges to x, and it is an accumulation point of E if and only if there is a net in $E \setminus \{x\}$ that converges to x.*

In a metric space, the cluster points of a sequence are precisely the limits of its convergent subsequences. An analogous statement holds for nets provided that one takes some care to define the notion of "subnet" properly. One might think that a subnet of $\langle x_\alpha \rangle_{\alpha \in A}$ should be simply the restriction of the map $\alpha \mapsto x_\alpha$ to a subset A' of A such that for every $\alpha \in A$ there is an $\alpha' \in A'$ with $\alpha \lesssim \alpha'$, but this is not general enough. Rather, a *subnet* of $\langle x_\alpha \rangle_{\alpha \in A}$ is defined to be the composition of the map $\alpha \mapsto x_\alpha$ with a map $\beta \mapsto \alpha_\beta$ from some other (perhaps more complicated) directed set B into A such that for every $\alpha_0 \in A$ there is a $\beta_0 \in B$ with $\alpha_0 \lesssim \alpha_\beta$ whenever $\beta_0 \lesssim \beta$; such a subnet is denoted by $\langle x_{\alpha_\beta} \rangle_{\beta \in B}$. With this definition, the cluster points of a net are precisely the limits of its convergent subnets.

1.4   COMPACTNESS

One is usually introduced to the term "compact" as a name for subsets of $\mathbb{R}^n$ that are both closed and bounded. When one generalizes to metric spaces, one quickly realizes that the real issue behind the closedness condition is completeness and that the boundedness condition must be replaced by something stronger. (An infinite set X equipped with the discrete metric where the distance between every two distinct points is 1 is bounded, but it is not compact in any reasonable sense.) The proper strengthening of boundedness turns out to be the following: a subset E of a metric space X is called *totally bounded* if for every $\epsilon > 0$, E is contained in the union of finitely many balls of radius ϵ. With this definition in hand, we have the following fundamental result.

1.10 Theorem. *If E is a subset of a metric space X, the following conditions are equivalent.*

   a. *E is complete and totally bounded.*
   b. *Every sequence in E has a subsequence that converges to a point of E.*
   c. *For every collection $\mathcal{U}$ of open sets in X such that $E \subset \bigcup_{U \in \mathcal{U}} U$ there is a finite subcollection $U_1, \ldots, U_n$ such that $E \subset \bigcup_1^n U_j$. (In brief: every open cover of E has a finite subcover.)*

The fact that (a) $\iff$ (b) is the **Bolzano-Weierstrass theorem**, and the fact that (a) $\iff$ (c) is the **Heine-Borel theorem**. A set E that possesses the properties (a)–(c) is called *compact*.

Now, how should this idea be generalized to topological spaces? Condition (a) does not make sense in that setting, and the remarks in the preceding section should lead one to suspect that condition (b) will contain some pitfalls. But condition (c) remains perfectly reasonable, and it turns out to be an extremely useful property, so it is taken as a definition. That is, a topological space X is *compact* if every open cover of X has a finite subcover; compactness for subsets of X means compactness in the relative topology. (The reader should be warned that other inequivalent definitions of compactness are sometimes encountered, particularly in the older literature.)

A useful equivalent formulation of compactness is obtained by considering the complements of the sets in an open cover and applying De Morgan's laws. It generalizes the nested interval theorem on the real line.

1.11 Proposition. *A topological space X is compact if and only if the following condition holds: whenever $\mathcal{E}$ is a family of closed subsets of X such that the intersection of every finite collection of sets in $\mathcal{E}$ is nonempty, then the intersection of all the sets in $\mathcal{E}$ is nonempty.*

Compactness is preserved under continuous mappings; that is, if $f : X \to Y$ is continuous and $E \subset X$ is compact, then so is $f(E)$. (The proof is easy: if $\{U_\alpha\}_{\alpha \in A}$ is an open cover of $f(E)$, then $\{f^{-1}(U_\alpha)\}_{\alpha \in A}$ is an open cover of E; take a finite subcover.) This is the abstract form of the extreme value theorem of elementary calculus.

The compactness condition achieves its maximum power in the setting of Hausdorff spaces. The following propositions set out the basic facts.

1.12 Proposition. *Suppose F is a compact subset of a Hausdorff space X. If $x \in X \setminus F$, there are disjoint open sets U and V such that $x \in U$ and $F \subset V$. Moreover, F is closed.*

1.13 Proposition. *Every compact Hausdorff space is normal.*

We sketch the proofs as a classic illustration of compactness arguments. For Proposition 1.12, since X is Hausdorff, for each $y \in F$ there are disjoint open sets U_y and V_y such that $x \in U_y$ and $y \in V_y$. The V_y's form an open cover of F, so there is a finite subcover $V_{y_1}, \ldots, V_{y_n}$, and we can take $U = \bigcap_1^n U_{y_j}$ and $V = \bigcup_1^n V_{y_j}$. This shows that each $x \in X \setminus F$ is an interior point of $X \setminus F$, so F is closed. As for Proposition 1.13, suppose X is compact Hausdorff and E and F are disjoint closed sets in X. By Proposition 1.12, for each $x \in E$ there are disjoint open sets U^x, V^x such that $x \in U^x$ and $F \subset V^x$. The U^x's form an open cover of E, so there is a finite subcover $U^{x_1}, \ldots, U^{x_m}$; then $U = \bigcup_1^m U^{x_j}$ and $V = \bigcap_1^m V^{x_j}$ are disjoint open sets containing E and F, respectively.

Let us re-examine condition (b) in Theorem 1.10. We can make it into a definition: A set E in a topological space is called *sequentially compact* if every sequence in E has a subsequence that converges to a point of E. In first countable spaces, compactness implies sequential compactness, but otherwise these two notions are unrelated in general. As one might expect, however, equivalence is restored by replacing sequences by nets. That is, a topological space X is compact if and only if every net in X has a convergent subnet.

How can one establish that a topological space is compact? For spaces of a finite-dimensional character (we are not being precise about what this means), compactness often follows from familiar facts about compact sets in Euclidean space together with the fact that the continuous image of a compact set is compact. For spaces of an infinite-dimensional character, however, compactness is a rather unusual phenomenon. The following theorem is one of the few powerful tools for obtaining it.

1.14 Tychonoff's theorem. *If $\{X_\alpha\}_{\alpha \in A}$ is an arbitrary family of compact topological spaces, then $\prod_{\alpha \in A} X_\alpha$, equipped with the product topology, is compact.*

There are several proofs of Tychonoff's theorem, all of which involve Zorn's lemma in one way or another. The one in [6, §4.6] is inspired by the following proof that the product of a finite collection $X_1, \ldots, X_k$ of compact metric spaces is compact. Suppose $\{x_n\}$ is a sequence in $\prod_1^k X_j$; thus $x_n = (x_n^1, \ldots, x_n^k)$ with $x_n^j \in X_j$. Choose a subsequence so that the first coordinates $\{x_n^1\}$ converge; then choose a subsequence of the latter so that the second coordinates converge, and so on by induction; eventually one obtains a convergent subsequence of the original sequence. This idea can be made to work in general: one replaces sequences by nets and the induction by an application of Zorn's lemma.

A topological space is called *locally compact* if every point has a compact neighborhood. Locally compact Hausdorff spaces (*LCH spaces*, for short) are an obviously useful class of spaces that include Euclidean spaces and their relatives. They are even more closely related to compact Hausdorff spaces than is immediately apparent; to wit, every noncompact LCH space can be compactified by adding a single "point at infinity."

More precisely, suppose X is a noncompact LCH space. Let $X^* = X \cup \{\infty\}$ where ∞ denotes a point that is not an element of X. We impose a topology on X^* by declaring a set $U \subset X^*$ to be open if either (i) U is an open subset of X, or (ii) $\infty \in U$ and $X^* \setminus U$ is a compact subset of X. It is easy to check that this defines a topology that makes X^* into a compact Hausdorff space, and also that the relative topology that X inherits as a subset of X^* is its original topology. The space X^* is called the *one-point compactification* of X.

For example, the one-point compactification of $\mathbb{R}^n$ is homeomorphic to the unit sphere in $\mathbb{R}^{n+1}$, $S^n = \{x \in \mathbb{R}^{n+1} : |x| = 1\}$. The easiest way to set this up is by stereographic projection. That is, think of $\mathbb{R}^n$ as sitting inside $\mathbb{R}^{n+1}$ as the set of points whose last coordinate is zero. For each $x \in \mathbb{R}^n$, draw the straight line from x to the "north pole" $\nu = (0, \ldots, 0, 1) \in S^n$, and let $f(x)$ be the point other than ν where this line intersects S^n. Then f is a homeomorphism from $\mathbb{R}^n$ to $S^n \setminus \{\nu\}$, and ν plays the role of the point at infinity.

By using the one-point compactification, one can easily transfer many results from compact Hausdorff spaces to LCH spaces. Most important, LCH spaces have a rich supply of continuous functions, and in particular of continuous functions that vanish outside compact sets. A bit of terminology: if f is a real- or complex-valued function on a topological space X, we define the *support* of f to be the closure of the set where f is nonzero and denote it by supp(f):

$$\mathrm{supp}(f) = \overline{\{x \in X : f(x) \neq 0\}}.$$

We then have the following version of Urysohn's lemma and the Tietze extension theorem, which is an easy consequence of Theorem 1.7.

1.15 Theorem. *Suppose X is an LCH space, K is a compact subset of X, and U is an open subset of X with $K \subset U$. If $\phi : K \to [a, b]$ is a continuous function on K, there is a continuous function $f : X \to [a, b]$ such that $f | K = \phi$ and $\mathrm{supp}(f)$ is compact and contained in U. In particular, there is a continuous $f : X \to [0, 1]$ such that $f = 1$ on K and $\mathrm{supp}(f)$ is compact and contained in U.*

We shall say more about continuous functions on LCH spaces in Chapter 5. We conclude this chapter with a result that has many interesting applications.

1.16 The Baire category theorem. *Suppose that X is either (i) a complete metric space or (ii) an LCH space. If $\{U_j\}_1^\infty$ is a countable collection of open dense sets in X, then $\bigcap_1^\infty U_j$ is dense in X.*

To prove this we must show that every nonempty open $W \subset X$ intersects $\bigcap_1^\infty U_j$. Since $W \cap U_1$ is open and nonempty, it contains a closed ball in case (i) and a compact set with nonempty interior in case (ii). Call the interior of this ball or compact set V_1. Since $V_1 \cap U_2$ is open and nonempty, it contains a closed ball in case (i) or a compact set with nonempty interior in case (ii); call the interior of this ball or compact set V_2. Continuing inductively, one obtains a nested sequence of closed balls or compact sets V_n such that $V_n \subset W \cap \bigcap_1^n U_j$. The radii of the balls may be taken to approach 0 so that their centers form a Cauchy sequence, so completeness in case (i) or Proposition 1.11 in case (ii) yields points in $W \cap \bigcap_1^\infty U_j$.

1.17 Corollary. *Suppose that X is a complete metric space or an LCH space. If $X = \bigcup_1^\infty F_j$ where each F_j is closed, then some F_j must have nonempty interior.*

This follows by applying the theorem to the sets $U_j = X \setminus F_j$. These open sets have empty intersection, so at least one of them must fail to be dense.

The word "category" in the name of Theorem 1.16 comes from Baire's original terminology, according to which a set is *of the first category* if it is a countable union of nowhere dense sets and *of the second category* otherwise. Corollary 1.17 thus says that complete metric spaces and LCH spaces are of the second category.

CHAPTER 2

MEASURE AND INTEGRATION: GENERAL THEORY

The theory of measure of subsets of Euclidean space (length, area, volume, and their analogues in higher dimensions) and the closely related theory of integration of functions on Euclidean space have a very long history. Much of the modern theory, however, does not depend on the particular features of the geometry of Euclidean space. It can be developed in a much more general setting with no additional effort, and in this more general form it yields results that can be applied in many additional situations.

This abstract theory is the subject of the present chapter; the methods for constructing interesting examples of measures and integrals and the study of the particular properties of these examples will be discussed in the following chapter. However, readers who prefer to anchor their thoughts in concrete situations are free to do so from the outset; we shall add a few comments at appropriate points for their benefit.

2.1 MEASURES

Roughly speaking, a measure on a space X is a function μ that assigns to a set $E \subset X$ a number $\mu(E) \in [0, \infty]$, such that $\mu(\bigcup_j E_j) = \sum_j \mu(E_j)$ whenever $\{E_j\}$ is a finite or infinite sequence of disjoint sets. The following are some typical situations that give rise to measures.

- If $X = \mathbb{R}^3$, $\mu(E)$ can be the volume of E.
- If X is a surface in $\mathbb{R}^3$, $\mu(E)$ can be the surface area of E.
- If X is a curve in $\mathbb{R}^3$, $\mu(E)$ can be the arc length of E.
- If we wish to model a distribution of mass in physical space, we can take $X = \mathbb{R}^3$ and $\mu(E)$ to be the amount of mass in E. This situation can

include features similar to all three of the preceding examples: distributions of mass with a continuous density throughout a 3-dimensional region, masses distributed over surfaces (thin plates), and masses distributed over curves (thin wires), as well as point masses (small massive objects).

- If X is the set of possible outcomes of a game or experiment with probabilistic features, $\mu(E)$ can represent the probability that the outcome lies in E.

At first one might think that the domain of a measure on X should be the collection of all subsets of X, but this is usually too much to ask for; it is often impossible to define measures of arbitrary sets in a way that is consistent with other features that one wants the measures to possess. Perhaps the most striking example of this sort of pathology is the following result, known as the **Banach-Tarski paradox**, a proof of which can be found in Stromberg [22].

2.1 Theorem. *Let U and V be bounded open sets in $\mathbb{R}^3$. There exist sets $E_1, \ldots, E_k$ and $F_1, \ldots, F_k$ such that:*
  a. $E_i \cap E_j = \varnothing$ and $F_i \cap F_j = \varnothing$ for all $i \neq j$.
  b. $U = \bigcup_1^k E_j$ and $V = \bigcup_1^k F_j$.
  c. For each j, F_j is the image of E_j under a rigid motion (i.e., a translation followed by a rotation).

Thus, for example, one can take a ball of radius 1, cut it up into a finite number of pieces, and rearrange the pieces to form two disjoint balls of radius 1. (The pieces are necessarily very bizarre; their existence depends on the axiom of choice.) This obviously precludes the existence of a notion of volume for arbitrary subsets of $\mathbb{R}^3$ such that the volume of a set is unchanged by rigid motions, as Euclidean geometry would require. The Banach-Tarski paradox is easily adapted to sets in $\mathbb{R}^n$ for any $n \geq 3$. It does not work in dimensions 1 or 2, but in $\mathbb{R}^1$ and $\mathbb{R}^2$ there are similar pathologies involving decompositions of a set into countably many pieces.

In view of these facts, one must take a little time to consider the families of sets that form appropriate domains for measures. Here is the standard terminology:

- We recall that the collection of all subsets of a set X is denoted by $\mathcal{P}(X)$.

- A nonempty family $\mathcal{A} \subset \mathcal{P}(X)$ is called an *algebra* if it is closed under finite unions and complements; that is, if $E_1, \ldots, E_n \in \mathcal{A}$ then $\bigcup_1^n E_j \in \mathcal{A}$, and if $E \in \mathcal{A}$ then $E^c = X \setminus E \in \mathcal{A}$.

- An algebra $A \subset \mathcal{P}(X)$ is called a *σ-algebra* if it is closed under countable unions; that is, if $E_1, E_2, \ldots \in A$ then $\bigcup_1^\infty E_j \in A$.

- A set X equipped with a σ-algebra $\mathcal{M} \subset \mathcal{P}(X)$ is called a *measurable space*.

If A is an algebra (resp. σ-algebra), then A is closed under finite (resp. countable) intersections, because $\bigcap_j E_j = \left[\bigcup_1^j E_j^c\right]^c$. Consequently, it is also closed under relative complements: if $E, F \in A$ then $E \setminus F = E \cap F^c \in A$. Moreover, if $A \subset \mathcal{P}(X)$ is an algebra, then $\varnothing \in A$ and $X \in A$, for if E is any element of A, then $\varnothing = E \cap E^c$ and $X = E \cup E^c$.

Measurable spaces bear a vague resemblance to topological spaces, and there are some structural similarities between the corresponding theories. In particular, if $(X, \mathcal{M})$ and $(Y, \mathcal{N})$ are measurable spaces, a map $f : X \to Y$ is called $(\mathcal{M}, \mathcal{N})$-*measurable*, or just *measurable* if $\mathcal{M}$ and $\mathcal{N}$ are understood, if $f^{-1}(E) \in \mathcal{M}$ for all $E \in \mathcal{N}$. Thus, the measurable maps are the analogues in the theory of measurable spaces of the continuous maps in the theory of topological spaces. However, it is generally much easier for two measurable spaces to be isomorphic (that is, for there to be a bijection f between the two spaces such that f and f^{-1} are both measurable) than it is for two topological spaces to be homeomorphic, so a σ-algebra on a space gives much less information about what the space really looks like than a topology does.

The most common way of producing σ-algebras is as follows. If $\mathcal{E}$ is any family of subsets of X, the intersection of all σ-algebras containing $\mathcal{E}$ (there is at least one, namely $\mathcal{P}(X)$) is again a σ-algebra; it is the smallest σ-algebra containing $\mathcal{E}$. It is called the σ-algebra *generated by* $\mathcal{E}$ and is denoted by $\mathcal{M}(\mathcal{E})$. Unlike the similar case of a topology generated by a family of sets, it is not easy to describe the elements of $\mathcal{M}(\mathcal{E})$ explicitly in terms of the elements of $\mathcal{E}$; passing from $\mathcal{E}$ to $\mathcal{M}(\mathcal{E})$ usually involves applying the operations of forming countable unions and complements infinitely many times.

There are two particularly important classes of examples of this construction. The first is the following. If X is a topological space, the σ-algebra generated by the family of open sets in X is called the *Borel σ-algebra* on X; it is denoted by $\mathcal{B}_X$, and its elements are called *Borel sets*. Thus $\mathcal{B}_X$ includes all open sets, closed sets, countable intersections of open sets (called G_δ *sets*), countable unions of closed sets (called F_σ *sets*), countable unions of G_δ sets (called $G_{\delta\sigma}$ *sets*), countable intersections of F_σ sets (called $F_{\sigma\delta}$ *sets*), and so forth.

The second important class of examples arises from Cartesian products. If $\{(X_\alpha, \mathcal{M}_\alpha)\}_{\alpha \in A}$ is a family of measurable spaces, the *product σ-algebra* on $X = \prod_{\alpha \in A} X_\alpha$ is the smallest σ-algebra on X that makes all the projection maps $\pi_\alpha : X \to X_\alpha$ measurable, that is, the σ-algebra generated by the sets $\pi_\alpha^{-1}(E_\alpha)$ as E_α ranges over $\mathcal{M}_\alpha$ and α ranges over A. It is denoted by $\bigotimes_{\alpha \in A} \mathcal{M}_\alpha$, or by $\mathcal{M}_1 \otimes \cdots \otimes \mathcal{M}_n$ if $A = \{1, \ldots, n\}$. (This is analogous to the product topology on a product of topological spaces.)

The interaction of these two constructions requires a little comment. Suppose that each X_α is a topological space. We can first form the Borel σ-algebras $\mathcal{B}_{X_\alpha}$ and then consider the product σ-algebra $\bigotimes_{\alpha \in A} \mathcal{B}_{X_\alpha}$, or we can first put the product topology on $X = \prod_{\alpha \in A} X_\alpha$ and then form its Borel σ-algebra $\mathcal{B}_X$. It is not hard to show that $\bigotimes_{\alpha \in A} \mathcal{B}_{X_\alpha}$ is generated by the sets $\pi_\alpha^{-1}(U_\alpha)$ where U_α is open in X_α and hence that

$$\bigotimes_{\alpha \in A} \mathcal{B}_{X_\alpha} \subset \mathcal{B}_X.$$

The reverse inclusion holds provided that each X_α is second countable and the index set A is countable — in particular, $\mathcal{B}_{\mathbb{R}^n} = \mathcal{B}_{\mathbb{R}} \otimes \cdots \otimes \mathcal{B}_{\mathbb{R}}$ — but otherwise it generally does not.

If $(X, \mathcal{M})$ and $(Y, \mathcal{N})$ are measurable spaces and $\mathcal{N}$ is generated by $\mathcal{E}$, for a map $f : X \to Y$ to be measurable it suffices to have $f^{-1}(E) \in \mathcal{M}$ for all $E \in \mathcal{E}$, because $\{E \subset Y : f^{-1}(E) \in \mathcal{M}\}$ is a σ-algebra on Y that contains $\mathcal{E}$ and hence contains $\mathcal{N}$. Two important cases:

- If X and Y are topological spaces, every continuous map $f : X \to Y$ is $(\mathcal{B}_X, \mathcal{B}_Y)$-measurable.

- A map f from a measurable space W into a product of measurable spaces $\prod_{\alpha \in A} X_\alpha$ (equipped with the product σ-algebra) is measurable if and only if $\pi_\alpha \circ f : W \to X_\alpha$ is measurable for all α.

For future reference, we make a few observations about the Borel σ-algebra on the real line. First, all intervals — open, closed, or half-open, and bounded or unbounded — are Borel sets. For open intervals this is true by definition of $\mathcal{B}_{\mathbb{R}}$, and we have, for example, $(a, b] = (a, \infty) \setminus (b, \infty)$ and $[a, b] = \mathbb{R} \setminus [(-\infty, a) \cup (b, \infty)]$. Second, since every open set in $\mathbb{R}$ is a countable union of open intervals, $\mathcal{B}_{\mathbb{R}}$ is generated by the open intervals. It is also generated by the bounded open intervals, the bounded closed intervals, the bounded half-open intervals open on the left or on the right, the open half-lines (a, ∞) or $(-\infty, a)$, or the corresponding closed half-lines. In all cases this is easily established by showing that all open in-

tervals can be obtained from the given family of sets by taking count-
able unions or intersections and complements — for example, $(a, b) = \bigcup_1^\infty [a + 2^{-n}, b - 2^{-n}]$.

With these preliminaries out of the way, we can give the formal definition of a measure. A *measure* on a measurable space $(X, \mathcal{M})$ is a map
$\mu : \mathcal{M} \to [0, \infty]$ such that

i. $\mu(\varnothing) = 0$.

ii. $\mu\left(\bigcup_1^\infty E_j\right) = \sum_1^\infty \mu(E_j)$ for every sequence $\{E_j\}_1^\infty$ of disjoint sets in $\mathcal{M}$.

Property (ii) is called *countable additivity*. Here are some more basic properties of measures:

- *Finite additivity*: If $\{E_j\}_1^n$ is a finite collection of disjoint sets in $\mathcal{M}$, then $\mu\left(\bigcup_1^n E_j\right) = \sum_1^n \mu(E_j)$.

- *Monotonicity*: If $E, F \in \mathcal{M}$ and $E \subset F$, then $\mu(E) \leq \mu(F)$.

- *Subadditivity*: If $\{E_j\}_1^\infty \subset \mathcal{M}$, then $\mu\left(\bigcup_1^\infty E_j\right) \leq \sum_1^\infty \mu(E_j)$.

- *Continuity from below*: If $\{E_j\} \subset \mathcal{M}$ and $E_j \subset E_{j+1}$ for all j, then $\mu\left(\bigcup_1^\infty E_j\right) = \lim_{j \to \infty} \mu(E_j)$.

- *Continuity from above*: If $\{E_j\}_1^\infty \subset \mathcal{M}$, $E_j \supset E_{j+1}$ for all j, and $\mu(E_1) < \infty$, then $\mu\left(\bigcap_1^\infty E_j\right) = \lim_{j \to \infty} \mu(E_j)$.

These are all easy to prove. Finite additivity follows from countable additivity because one can take $E_j = \varnothing$ for $j > n$, and continuity from below
holds because

$$\mu\left(\bigcup_1^\infty E_j\right) = \sum_1^\infty \mu(E_j \setminus E_{j-1}) = \lim_{n \to \infty} \sum_1^n \mu(E_j \setminus E_{j-1}) = \lim_{n \to \infty} \mu(E_n).$$

The other assertions follow by similar arguments.

A measurable space $(X, \mathcal{M})$ equipped with a measure μ is called a *measure space*. The measure space $(X, \mathcal{M}, \mu)$, or just the measure μ, is called
finite if $\mu(X) < \infty$, and it is called *σ-finite* if $X = \bigcup_1^\infty E_j$ where $E_j \in \mathcal{M}$
and $\mu(E_j) < \infty$ for all j. The condition of σ-finiteness is satisfied by
the vast majority of measures that turn up in practice, and it precludes certain pathologies that make the theory of arbitrary measures more complicated. We shall therefore assume that all measures in question are σ-finite
whenever that simplifies the discussion; the reader is referred to [6] for the
full story.

If X is a topological space and $\mathcal{B}_X$ is its Borel algebra, a measure on
$(X, \mathcal{B}_X)$ is called a *Borel measure*.

If $(X, \mathcal{M}, \mu)$ is a measure space, a set $E \in \mathcal{M}$ such that $\mu(E) = 0$ is called a *null set*, or a *μ-null set* when the added precision is necessary. A property of points $x \in X$ that holds except on a null set is said to hold *almost everywhere* or *for almost every x* or just *a.e.*; again, the prefix "μ-" can be added as needed. (Probabilists generally use the term "almost surely" instead of "almost everywhere.")

If $\mu(E) = 0$, then $\mu(F) = 0$ for all $F \in \mathcal{M}$ such that $F \subset E$. In general there is no reason for arbitrary subsets of null sets to belong to $\mathcal{M}$, but it is often convenient to assume that they do — a condition known as *completeness* of μ — and it is always easy to arrange this by enlarging the domain of μ. More precisely, let

$$\overline{\mathcal{M}} = \{E \cup F : E \in \mathcal{M} \text{ and } F \subset N \text{ where } N \in \mathcal{M} \text{ and } \mu(N) = 0\},$$

and define $\overline{\mu} : \overline{\mathcal{M}} \to [0, \infty]$ by setting $\overline{\mu}(E \cup F) = \mu(E)$. It is then easy to verify that $\overline{\mathcal{M}}$ is a σ-algebra, that $\overline{\mu}$ is indeed well defined (i.e., if $E \cup F = E' \cup F'$ where the E's and F's are as above, then $\mu(E) = \mu(E')$), and that $\overline{\mu}$ is a complete measure on $\overline{\mathcal{M}}$ that agrees with μ on $\mathcal{M}$. $\overline{\mu}$ is called the *completion* of μ.

The construction of useful measures is a nontrivial task to which we shall turn in Chapter 3. For the present, we briefly mention a couple of easy examples and the classic not-so-easy examples so that the reader may have some concrete points of reference.

- Let $(X, \mathcal{M})$ be a measurable space. *Counting measure* is the measure μ on $\mathcal{M}$ such that $\mu(E)$ is the number of elements of E whenever E is finite and $\mu(E) = \infty$ otherwise. If $x \in X$, the *point mass at x* is the measure δ_x on $\mathcal{M}$ such that $\delta_x(E) = 1$ if $x \in E$ and $\delta_x(E) = 0$ otherwise.

- There is a unique measure on $(\mathbb{R}, \mathcal{B}_{\mathbb{R}})$ such that the measure of each interval is its length, and for $n > 1$ there is a unique measure on $(\mathbb{R}^n, \mathcal{B}_{\mathbb{R}^n})$ such that the measure of the Cartesian product of n intervals is the product of their lengths. The completions of these measures are called *Lebesgue measure* on $\mathbb{R}$ and $\mathbb{R}^n$, respectively.

2.2 INTEGRATION

In this section we develop the theory of integration of real- or complex-valued functions on a measure space. The basic building blocks of this theory are the so-called "simple functions." Here are the definitions.

- If X is any nonempty set and $E \subset X$, the *characteristic function* or *indicator function* of E is the function $\chi_E : X \rightarrow \{0, 1\}$ (sometimes denoted by 1_E) defined by

$$\chi_E(x) = \begin{cases} 1 & \text{if } x \in E, \\ 0 & \text{if } x \notin E. \end{cases}$$

- If $(X, \mathcal{M})$ is a measurable space, a *simple function* on X is a finite linear combination, with complex coefficients, of characteristic functions of measurable sets.

Every simple function can be written uniquely as $\sum_1^n c_j \chi_{E_j}$ where the c_j's are distinct complex numbers (one of which may be 0) and the E_j's are disjoint measurable sets whose union is X.

Now suppose $(X, \mathcal{M}, \mu)$ is a measure space. If $\phi = \sum_1^n c_j \chi_{E_j}$ is a non-negative simple function, its integral with respect to μ, $\int \phi \, d\mu$, is defined in the obvious way:

$$\int \phi \, d\mu = \sum c_j \mu(E_j),$$

with the understanding that if $c_j = 0$ and $\mu(E_j) = \infty$ then $c_j \mu(E_j) = 0$. Note that $\int \phi \, d\mu$ may be $+\infty$ if some of the sets E_j have infinite measure.

To extend this notion of integral to more general functions, we approximate such functions by simple functions. To fix the ideas, suppose f is a bounded nonnegative function on X, say $f : X \rightarrow [0, b)$. (It is convenient to use half-open intervals since they fit together without overlapping.) We can approximate f by a simple function by cutting up the interval $[0, b)$ into 2^n equal subintervals $[0, b/2^n)$, $[b/2^n, 2b/2^n)$, etc., and replacing $f(x)$ by the constant $jb/2^n$ on the set where $jb/2^n \le f(x) < (j+1)b/2^n$. That is, let

$$(2.2) \qquad E_j^n = f^{-1}\left(\left[\frac{jb}{2^n}, \frac{(j+1)b}{2^n}\right)\right), \qquad \phi_n = \sum_{j=0}^{2^n-1} \frac{jb}{2^n} \chi_{E_j^n}.$$

Then, since $E_j^n = E_{2j}^{n+1} \cup E_{2j+1}^{n+1}$ for all j and n, the ϕ_n's form an increasing sequence of functions that converges uniformly to f as $n \rightarrow \infty$; in fact, $0 \le f(x) - \phi_n(x) < 1/2^n$ for all $x \in X$.

We would like to take $\int f \, d\mu$ to be the limit of the Riemann-type sums $\int \phi_n \, d\mu$. For this to work, however, the sets E_j must be measurable. In other words, we need to assume that $f^{-1}(I)$ is in $\mathcal{M}$ for any half-open interval

I. Since these intervals generate the Borel σ-algebra $\mathcal{B}_{\mathbb{R}}$, the condition we want is that f be $(\mathcal{M}, \mathcal{B}_{\mathbb{R}})$-measurable.

A definition is in order: If $(X, \mathcal{M})$ is a measurable space, a real-valued function f on X is said to be *measurable* (with no further qualification) if it is $(\mathcal{M}, \mathcal{B}_{\mathbb{R}})$-measurable, and a complex-valued function f on X is said to be *measurable* if its real and imaginary parts are measurable, or equivalently if it is $(\mathcal{M}, \mathcal{B}_{\mathbb{C}})$-measurable. Sometimes it is convenient to consider functions with values in $[-\infty, \infty]$; such a function f will be called *measurable* if $f^{-1}(I) \in \mathcal{M}$ for any interval $I \subset [-\infty, \infty]$ (whose endpoints might be infinite). Finally, if $E \in \mathcal{M}$, a function on E is said to be *measurable* if f is measurable when extended to X by the prescription $f(x) = 0$ for $x \notin E$.

The class of measurable functions is preserved by most of the common operations on functions. In particular, we have:

2.3 Proposition. *Let $(X, \mathcal{M})$ be a measurable space.*

a. *If $f : X \to \mathbb{R}$ is measurable and $\phi : \mathbb{R} \to \mathbb{R}$ is continuous, then $\phi \circ f$ is measurable.*

b. *If $f, g : X \to \mathbb{R}$ are measurable, then so are $f + g$ and fg.*

c. *If $f_1, f_2, \ldots : X \to [-\infty, \infty]$ are measurable, then so are $\sup_j f_j$, $\inf_j f_j$, $\limsup_{j \to \infty} f_j$, and $\liminf_{j \to \infty} f_j$ (all of these operations being defined pointwise).*

d. *If $f_1, f_2, \ldots : X \to [-\infty, \infty]$ are measurable and the limit $f(x) = \lim_{n \to \infty} f_n(x)$ exists for all $x \in X$, then f is measurable.*

Assertion (a) is obvious since $(\phi \circ f)^{-1}(E) = f^{-1}(\phi^{-1}(E))$ and continuous functions are $(\mathcal{B}_{\mathbb{R}}, \mathcal{B}_{\mathbb{R}})$-measurable. Likewise, (b) follows from the fact that $f + g$ and fg are the compositions of the measurable map $(f, g) : X \to \mathbb{R}^2$ with the continuous maps $(s, t) \mapsto s + t$ and $(s, t) \mapsto st$ from $\mathbb{R}^2$ to $\mathbb{R}$. For (c), we use the facts that $(\sup_j f_j)^{-1}((a, \infty]) = \bigcup_1^\infty f_j^{-1}((a, \infty])$ (and similarly for $\inf_j f_j$) and $\limsup_j f_j = \inf_{k \geq 1}[\sup_{j \geq k} f_j]$ (and similarly for $\liminf f_j$). Finally, (d) is a corollary of (c).

Parts (a), (b), and (d) of Proposition 2.3 have obvious analogues for complex-valued functions. The first two of these are proved in the same way as above, and the last follows by considering real and imaginary parts separately. By the way, let us underline the power of part (d): when $X = [a, b]$, the analogue of (d) with "measurable" replaced by "Riemann integrable" is false!

Here are two simple and useful decompositions of functions. First, if $f : X \to [-\infty, \infty]$, we define the *positive* and *negative parts* of f to be

$$(2.4) \qquad f^+(x) = \max\left(f(x), 0\right), \qquad f^-(x) = \max\left(-f(x), 0\right),$$

so that $f = f^+ - f^-$. These are measurable if f is, by Proposition 2.3(c). Second, if $f : X \to \mathbb{C}$, the *polar decomposition* of f is

$$(2.5) \qquad f = (\operatorname{sgn} f)|f|, \quad \text{where} \quad \operatorname{sgn} z = \begin{cases} z/|z| & \text{if } z \neq 0, \\ 0 & \text{if } z = 0. \end{cases}$$

Again, if f is measurable, so are $\operatorname{sgn} f$ and $|f|$, since the absolute value function is continuous and the signum function is continuous except at the origin.

Arbitrary measurable functions f can be approximated by simple functions. We showed how to do this when f is bounded and nonnegative in (2.2). If f is nonnegative but unbounded, one can apply this construction to the function $g^m(x) = \min(f(x), m)$ to obtain a sequence of simple functions $\{\phi_n^m\}_{n=1}^\infty$ that increases uniformly to g^m, and then take $\phi_n = \phi_n^n$ to obtain a sequence that increases pointwise to f. If $f : X \to [-\infty, \infty]$, we can apply this construction to the positive and negative parts of f, and if $f : X \to \mathbb{C}$, we can work with the real and imaginary parts of f. The upshot is the following result.

2.6 Proposition. *If f is a measurable function on X with values in $\mathbb{R}$, $\mathbb{C}$, or $[-\infty, \infty]$, there is a sequence $\{\phi_n\}$ of simple functions, with $\phi_n \geq 0$ if $f \geq 0$, such that $\phi_n \to f$ pointwise and $|\phi_n|$ increases to $|f|$ pointwise; moreover, $\phi_n \to f$ uniformly if f is bounded.*

We are now ready to define the integral in general. Given a measure space $(X, \mathcal{M}, \mu)$, we set

$$L^+(X) = \{f : X \to [0, \infty] : f \text{ is measurable}\},$$

and for $f \in L^+(X)$ we define the *integral of f with respect to μ* by

$$\int f \, d\mu = \sup \left\{ \int \phi \, d\mu : \phi \text{ is simple and } 0 \leq \phi \leq f \right\}.$$

Thus $\int f \, d\mu$ is an element of $[0, \infty]$. The integral is *monotonic*: if $f, g \in L^+(X)$, then clearly

$$f \leq g \quad \Longrightarrow \quad \int f \, d\mu \leq \int g \, d\mu.$$

Next, we say that a measurable $f : X \to \mathbb{C}$ is *integrable* if $\int |f| \, d\mu < \infty$, and we denote the set of integrable functions by $L^1(X, \mu)$:

$$L^1(X, \mu) = \left\{ f : X \to \mathbb{C} : f \text{ is measurable and } \int |f| \, d\mu < \infty \right\}.$$

Given a measurable $f : X \to \mathbb{C}$, let $g = \operatorname{Re} f$ and $h = \operatorname{Im} f$. We observe that f is integrable if and only if the integrals of $g^{\pm}$ and $h^{\pm}$ are all finite; this follows from monotonicity in view of the inequalities

$$g^{\pm} \leq |f|, \qquad h^{\pm} \leq |f|, \qquad |f| \leq g^{+} + g^{-} + h^{+} + h^{-}.$$

For $f \in L^1(X, \mu)$, we may therefore define the *integral of* $f = g + ih$ *with respect to* μ to be

$$\int f \, d\mu = \int g^{+} \, d\mu - \int g^{-} \, d\mu + i \int h^{+} \, d\mu - i \int h^{-} \, d\mu.$$

Some matters of notation: We often write $L^1(\mu)$, $L^1(X)$, or just L^1 for $L^1(X, \mu)$ when the meaning is clear from the context; likewise L^{+} for $L^{+}(X)$. When μ is understood, we may write $\int f$ for $\int f \, d\mu$; on the other hand, when it is convenient to display the argument of f explicitly (for example, because there are other variables present), we may write $\int f(x) \, d\mu(x)$ instead. Moreover, if E is a measurable set in X, we define the integral of f over E by

$$\int_{E} f \, d\mu = \int f \chi_{E} \, d\mu.$$

We summarize the basic properties of the integral in a theorem.

2.7 Theorem. *Let $(X, \mathcal{M}, \mu)$ be a measure space.*
   a. *L^1 is a vector space, and the integral $f \mapsto \int f$ is a linear functional on it.*
   b. *The linearity conditions $\int (f + g) = \int f + \int g$ and $\int cf = c \int f$ hold also for $f, g \in L^{+}$ when $c > 0$.*
   c. *If $f \in L^1$, then $|\int f| \leq \int |f|$.*
   d. *If $f, g \in L^1$ or $f, g \in L^{+}$, then $\int_{E} f = \int_{E} g$ for all measurable $E \subset X$ if and only if $\int |f - g| = 0$ if and only if $f = g$ μ-almost everywhere.*

Most of these assertions follow easily from the definitions; the one that takes some work is additivity. One first shows that $\int (f + g) = \int f + \int g$ when f and g are simple by a direct calculation, then when f and g are in L^{+} by using Proposition 2.6 together with the monotone convergence theorem (which we present in the next section), and finally for $f, g \in L^1$ by reducing to the case of nonnegative functions.

Theorem 2.7(d) shows that as far as integration is concerned, there is no difference between a function f and any other function that is equal to f almost everywhere. In fact, we can integrate functions that are only defined almost everywhere provided that they can be extended to the whole space in a measurable way (say, by setting them equal to 0 where they are not already defined); this is a convenience in dealing with functions with discontinuities and other singularities.

To put this another way, the set N of measurable functions that are equal to 0 almost everywhere is a vector subspace of L^1, and the integral is really a linear functional on the quotient space $\widetilde{L}^1 = L^1/N$. The modified space $\widetilde{L}^1$ has the advantage that the functional

$$(2.8) \qquad \rho(f, g) = \int |f - g|$$

is a metric on it. (The triangle inequality is true already in L^1, but one has to pass to $\widetilde{L}^1$ to obtain the condition that $\rho(f, g) = 0$ only when $f = g$.) Moreover, on this level it makes little difference whether the measure μ is complete or not. Indeed, if $(X, \overline{\mathcal{M}}, \overline{\mu})$ is the completion of $(X, \mathcal{M}, \mu)$, every $\overline{\mathcal{M}}$-measurable function agrees $\overline{\mu}$-almost everywhere with an $\mathcal{M}$-measurable function, so that $\widetilde{L}^1(\overline{\mu}) = \widetilde{L}^1(\mu)$.

By common agreement, however, no notational distinction is made between L^1 and $\widetilde{L}^1$. That is, both of these spaces are denoted by L^1, and when we write "$f \in L^1$" we may understand f either as an integrable function or as an equivalence class of such functions, the equivalence being equality almost everywhere, depending on the context. This almost never causes any confusion.

2.3 CONVERGENCE OF FUNCTIONS AND CONVERGENCE OF INTEGRALS

In this section we work on a fixed measure space $(X, \mathcal{M}, \mu)$. We begin with a group of fundamental theorems that address the question of when "the integral of the limit is the limit of the integrals." The first one pertains to functions in L^+.

2.9 The monotone convergence theorem. *Let $\{f_n\}$ be a sequence in L^+ such that $f_n(x) \leq f_{n+1}(x)$ for all n and x, and let*

$$f(x) = \lim_{n \to \infty} f_n(x) = \sup_n f_n(x)$$

(which always exists since we allow the value ∞). Then

$$\int f = \lim_{n \to \infty} \int f_n.$$

The hypothesis that $f_n \leq f_{n+1}$ cannot be omitted. Consider the following three sequences of functions on $\mathbb{R}$:

$$(2.10) \qquad f_n = n\chi_{(0,1/n)}, \qquad g_n = \chi_{(n,n+1)}, \qquad h_n = n^{-1}\chi_{(0,n)}.$$

These sequences all converge to 0 pointwise, and the last one even converges uniformly, but the integral of each of these functions with respect to Lebesgue measure is equal to 1. In all cases, the trouble is that the area under the graph "leaks out to infinity" as $n \to \infty$, so the integral of the limit is less than one might expect. This suggests that there might be an inequality that is valid more generally, and that is the case.

2.11 Fatou's lemma. *If $\{f_n\}$ is any sequence in L^+, then*

$$\int (\liminf_{n\to\infty} f_n) \leq \liminf_{n\to\infty} \int f_n.$$

In particular, if $f_n \to f$ almost everywhere, then $\int f \leq \liminf_{n\to\infty} \int f_n$.

Finally, if we impose a bound on the functions f_n that forbids the areas under their graphs to escape to infinity, we get another positive result.

2.12 The dominated convergence theorem. *If $\{f_n\}$ is a sequence in L^1 such that $f_n \to f$ almost everywhere, and there exists $g \in L^+ \cap L^1$ such that $|f_n| \leq g$ almost everywhere for all n, then*

$$\int f = \lim_{n\to\infty} \int f_n.$$

As for the proofs: In the monotone convergence theorem, $\{\int f_n\}$ is clearly an increasing sequence whose limit is at most $\int f$, by monotonicity; the proof of the reverse inequality takes some ingenuity. Fatou's lemma is a rather easy consequence of the monotone convergence theorem and the fact that $\liminf f_n = \sup_{k\geq 1}(\inf_{n\geq k} f_n)$. The dominated convergence theorem is proved, for real-valued f_n, by applying Fatou's lemma to the sequences $\{g + f_n\}$ and $\{g - f_n\}$; the complex case follows immediately.

These theorems about sequences can immediately be translated into theorems about infinite series by applying them to the partial sums of the series. In particular, if $\{f_n\}$ is any sequence in L^+, the monotone convergence theorem implies that $\int \sum_n f_n = \sum_n \int f_n$. In the case of L^1 functions there is a bit more to be said; the point is that the finiteness of $\sum_n \int |f_n|$ implies the finiteness of $\sum_n |f_n(x)|$ for almost every x, which in turn implies the convergence of $\sum_n f_n(x)$ for almost every x.

2.13 Theorem. *Suppose that* $\{f_n\}$ *is a sequence in* L^1 *such that* $\sum_n \int |f_n|$ $< \infty$. *Then the series* $\sum_n f_n$ *converges almost everywhere to a function* $f \in L^1$, *and* $\int f = \sum_n \int f_n$.

Another use of the dominated convergence theorem is in obtaining sharp forms of some theorems from advanced calculus concerning integrals containing parameters.

2.14 Theorem. *Given a measure space* $(X, \mathcal{M}, \mu)$ *and a complex-valued function* f *on* $X \times [a, b]$ *such that* $f(\cdot, t) \in L^1(\mu)$ *for each* $t \in [a, b]$, *let* $F(t) = \int_X f(x, t) \, d\mu(x)$.
  a. *Suppose there is a* $g \in L^1(\mu)$ *such that* $|f(x, t)| \le g(x)$ *for all* x, t. *If* $\lim_{t \to t_0} f(x, t) = f(x, t_0)$ *for every* x, *then* $\lim_{t \to t_0} F(t) = F(t_0)$. *In particular, if* $f(x, \cdot)$ *is continuous for each* x, *then* F *is continuous.*
  b. *Suppose that the partial derivative* $\partial_t f = \partial f / \partial t$ *exists and there is a* $g \in L^1(\mu)$ *such that* $|\partial_t f(x, t)| \le g(x)$ *for all* x, t. *Then* F *is differentiable and* $F'(t) = \int \partial_t f(x, t) \, d\mu(x)$.

To prove part (b), one applies the dominated convergence theorem to the difference quotients $[F(t) - F(t_0)]/(t - t_0)$, where t is constrained to approach t_0 through a sequence of values; the required domination of these quotients follows from the estimate for $|\partial_t f|$ by the mean value theorem.

We now turn to a comparison of different modes of convergence. If $\{f_n\}$ is a sequence of functions on a set X, the statement "$f_n \to f$" can mean many things. The meaning that one usually meets first is *pointwise* convergence ($f_n(x) \to f(x)$ for every x), and advanced calculus books stress the importance of *uniform* convergence ($\sup_x |f_n(x) - f(x)| \to 0$). On a measure space, we can now add *convergence almost everywhere* ($f_n(x) \to f(x)$ for almost every x) and *convergence in* L^1 ($\int |f_n - f| \to 0$, i.e., convergence with respect to the metric (2.8)). There is one more useful notion that can be added to this list: we say that $f_n \to f$ *in measure* if for every $\epsilon > 0$ we have

$$\mu\left(\{x : |f_n(x) - f(x)| \ge \epsilon\}\right) \to 0 \text{ as } n \to \infty.$$

Of course uniform convergence $\Rightarrow$ pointwise convergence $\Rightarrow$ convergence almost everywhere, but not conversely (except in special cases). The sequences in (2.10) show that pointwise convergence or even uniform convergence does not imply convergence in L^1; the first and third also show that convergence in measure does not imply convergence in L^1, and the second shows that pointwise convergence does not imply convergence in measure. Moreover, it is not hard to construct a sequence $\{f_n\}$ that converges in L^1 and in measure but such that $\{f_n(x)\}$ does not converge for any x. (Take $X = [0, 1]$ with Lebesgue measure, and take the functions f_n to be the characteristic functions of intervals whose lengths tend to 0 but whose midpoints oscillate back and forth over $[0, 1]$ so that for each x there are infinitely many n with $f_n(x) = 1$ and infinitely many n with $f_n(x) = 0$.)

After this avalanche of negative results, here is a summary of the positive ones.

2.15 Theorem.
a. *Uniform convergence and convergence in L^1 each imply convergence in measure.*
b. *If $f_n \to f$ almost everywhere and there exists $g \in L^+ \cap L^1$ such that $|f_n| \le g$ for all n, then $f_n \to f$ in L^1.*
c. *If $f_n \to f$ in measure, there is a subsequence $\{f_{n_j}\}$ that converges to f almost everywhere.*
d. *If $\mu(X) < \infty$, convergence almost everywhere implies convergence in measure.*

The first assertion is an easy exercise, and the second follows from the dominated convergence theorem. The other two require some work to prove.

2.4 PRODUCT MEASURES AND THE FUBINI-TONELLI THEOREM

Suppose that $(X_j, \mathcal{M}_j, \mu_j)$ is a σ-finite measure space for $j = 1, \ldots, n$, and let $X = \prod_1^n X_j$ and $\mathcal{M} = \bigotimes_1^n \mathcal{M}_j$. By an argument that we shall sketch at the end of §3.1, there is a unique measure π on $(X, \mathcal{M})$ such that

$$\pi(E_1 \times E_2 \times \cdots \times E_n) = \mu_1(E_1)\mu_2(E_2)\cdots\mu_n(E_n) \text{ for all } E_j \in \mathcal{M}_j,$$

with the understanding that any numerical product containing 0 as a factor has the value 0, even if one or more of the other factors is ∞. This measure is called the *product* of $\mu_1, \ldots, \mu_n$ and is denoted by $\mu_1 \times \cdots \times \mu_n$. (An analogous result holds even for infinitely many factors provided that $\mu_j(X_j) = 1$ for all but finitely many j, a useful result in probability theory. A nice proof can be found in Saeki [19].) In what follows we restrict the discussion to the case of two factors to keep the notation more manageable, but the generalization to n factors is straightforward.

Suppose then that $(X, \mathcal{M}, \mu)$ and $(Y, \mathcal{N}, \nu)$ are σ-finite measure spaces. If f is a function on $X \times Y$, we can consider not only the integral of f with respect to the product measure but also the iterated integrals of f with respect to μ and ν or with respect to ν and μ. It will be convenient to employ the following notation for the functions on X and Y obtained from f by fixing one of its arguments:

$$f^y(x) = f(x, y) = f_x(y).$$

Here is the main result. Parts (b) and (c) are due to Tonelli and Fubini, respectively, in the case where $X = Y = \mathbb{R}$ and $\mu = \nu = $ Lebesgue measure. Fubini came first, and the whole theorem is often called simply **Fubini's theorem.**

2.16 The Fubini-Tonelli theorem. *Let $(X, \mathcal{M}, \mu)$ and $(Y, \mathcal{N}, \nu)$ be σ-finite measure spaces.*

   a. *If f is an $\mathcal{M} \otimes \mathcal{N}$-measurable function on $X \times Y$, then f^y is $\mathcal{M}$-measurable for all $y \in Y$ and f_x is $\mathcal{N}$-measurable for all $x \in X$.*
   b. *If $f \in L^+(X \times Y)$, the functions $g(x) = \int f_x \, d\nu$ and $h(y) = \int f^y \, d\mu$ are in $L^+(X)$ and $L^+(Y)$, respectively, and*

$$\int_{X \times Y} f \, d(\mu \times \nu) = \int_X \left[\int_Y f(x, y) \, d\nu(y) \right] d\mu(x)$$

(2.17)

$$= \int_Y \left[\int_X f(x, y) \, d\mu(x) \right] d\nu(y).$$

   c. *If $f \in L^1(X \times Y)$, then $f_x \in L^1(\nu)$ for almost every $x \in X$ and $f^y \in L^1(\mu)$ for almost every $y \in Y$; the almost-everywhere-defined functions $g(x) = \int f_x \, d\nu$ and $h(y) = \int f^y \, d\mu$ are in $L^1(\mu)$ and $L^1(\nu)$, respectively; and (2.17) holds.*

Part (a) is an easy exercise. The hard part of the proof is establishing part (b) when f is the characteristic function of a set in $\mathcal{M} \otimes \mathcal{N}$. Once this is done, (b) follows in general by a limiting argument involving the monotone convergence theorem, and (c) is an easy corollary.

It should be noted that the measure $\mu \times \nu$ is usually not complete even when μ and ν are complete. (For example, if E is a nonmeasurable set in X and N is a nonempty set of measure zero in Y, then $E \times N \notin \mathcal{M} \otimes \mathcal{N}$, but $E \times N$ is a subset of $X \times N$, a set of measure zero in $X \times Y$.) For some purposes it is preferable to state the Fubini-Tonelli theorem in a way that involves the completion λ of $\mu \times \nu$. This is easy to arrange. In the statement of the theorem, one must simply add the assumption that μ and ν are complete, replace "all" by "almost all" in two places in part (a), and replace $\mu \times \nu$ by λ in (2.17). The functions g and h are then defined almost everywhere in (b) as well as in (c), and the rest of the theorem remains valid as stated.

The Fubini-Tonelli theorem is an essential tool in analysis. It is most commonly used to justify interchanging the order of integration in an iterated integral, and the parts (b) and (c) of the theorem typically function as a team: First one verifies that $f \in L^1(\mu \times \nu)$ by using part (b) to evaluate $\int |f| \, d(\mu \times \nu)$ as an iterated integral in one order or the other; then one invokes part (c) to effect the desired interchange. (Incidentally, Theorem 2.13 is essentially the special case of Fubini's theorem where one of the factors is the set of positive integers with counting measure.)

2.5   RELATIONS BETWEEN (SIGNED AND COMPLEX) MEASURES

Suppose μ and ν are two measures on the same measurable space $(X, \mathcal{M})$. We say that μ and ν are *mutually singular* and write $\mu \perp \nu$ if there are sets E and F in $\mathcal{M}$ such that

$$E \cap F = \varnothing, \quad E \cup F = X, \quad \nu(E) = \mu(F) = 0.$$

(We also say that ν is *singular with respect to* μ or vice versa.) Thus mutual singularity means that μ and ν "live on different parts of X." (These two parts E and F may not be separated in a clean geometric way. For example, on $(\mathbb{R}, \mathcal{B}_{\mathbb{R}})$, let μ be Lebesgue measure and let ν be counting measure on the rationals — that is, for $A \in \mathcal{B}_{\mathbb{R}}$, $\nu(A)$ is the number of rational points in A. Then $\mu \perp \nu$, as one can take E to be the set of irrationals and F the set of rationals.)

On the other hand, we say that ν is *absolutely continuous with respect to* μ and write $\nu \ll \mu$ if $\nu(E) = 0$ for every $E \in \mathcal{M}$ such that $\mu(E) = 0$. (Thus "ν lives on the same part of X as μ.") Absolute continuity and mutual singularity are essentially mutually exclusive: if $\nu \perp \mu$ and $\nu \ll \mu$ then $\nu = 0$. (The term "absolutely continuous" derives from an older terminology for functions on $\mathbb{R}$ that we shall meet in §3.3.) That absolute continuity is indeed a form of continuity is indicated by the following result:

2.18 Theorem. *Suppose that μ and ν are measures on $(X, \mathcal{M})$ and ν is finite. Then $\nu \ll \mu$ if and only if for every $\epsilon > 0$ there is a $\delta > 0$ such that $\nu(E) < \epsilon$ for every $E \in \mathcal{M}$ such that $\mu(E) < \delta$.*

Examples of absolute continuity are easily generated as follows: given $f \in L^+(\mu)$, define ν on $\mathcal{M}$ by $\nu(E) = \int_E f \, d\mu$. The finite additivity of ν follows from the linearity of the integral, and countable additivity then follows from the monotone convergence theorem, so ν is indeed a measure, and it is obvious that $\nu \ll \mu$. Moreover, it is easily checked that $\int g \, d\nu = \int fg \, d\mu$ for all $g \in L^+(X)$, by approximating g by simple functions. We may therefore indicate the relation between μ, ν, and f briefly as

$$d\nu = f \, d\mu,$$

and we may bend the language a bit by speaking of "the measure $f \, d\mu$." The following fundamental theorem shows, among other things, that these examples are essentially the only ones.

2.19 The Lebesgue-Radon-Nikodym theorem. *Suppose μ and ν are σ-finite measures on $(X, \mathcal{M})$. There exist unique σ-finite measures ν_{ac} and ν_s on $(X, \mathcal{M})$ such that*

$$\nu = \nu_{ac} + \nu_s, \qquad \nu_{ac} \ll \mu, \qquad \nu_s \perp \mu.$$

Moreover, there exists $f \in L^+(X)$ such that $d\nu_{ac} = f \, d\mu$, and any two such functions are equal μ-almost everywhere.

When μ and ν are both finite, one can prove this theorem by showing that among all functions $g \in L^+(X)$ such that $\int_E g \, d\mu \leq \nu(E)$ for all $E \in \mathcal{M}$ there is one such that $\int_X g \, d\mu$ is maximal. Taking f to be this function, one defines ν_{ac} and ν_s by $d\nu_{ac} = f \, d\mu$ and $\nu_s = \nu - \nu_{ac}$, then shows that $\nu_s \perp \mu$. The σ-finite case follows by decomposing X into a countable union of sets E_j such that $\mu(E_j)$ and $\nu(E_j)$ are both finite and applying this argument on each E_j.

The decomposition $\nu = \nu_{ac} + \nu_s$ is called the *Lebesgue decomposition* of ν with respect to μ. When $\nu \ll \mu$, so that $\nu = \nu_{ac}$, the function f is called the *Radon-Nikodym derivative* of ν with respect to μ and is denoted by $d\nu/d\mu$; the fact that it exists is the **Radon-Nikodym theorem**. The chain rule for Radon-Nikodym derivatives is easy to verify: if $\nu \ll \mu$ and $\mu \ll \lambda$, then $\nu \ll \lambda$ and

$$\frac{d\nu}{d\lambda} = \frac{d\nu}{d\mu} \frac{d\mu}{d\lambda} \qquad \lambda\text{-almost everywhere.}$$

Much of the preceding material can be generalized from measures to countably additive set functions whose values are not necessarily nonnegative, and this will be significant in the connections with other subjects that we shall discuss in §3.3 and §5.2. Here are the definitions: A *signed measure* (resp. *complex measure*) on a measurable space $(X, \mathcal{M})$ is a map ν from $\mathcal{M}$ to either $(-\infty, \infty]$ or $[-\infty, \infty)$ (resp. $\mathbb{C}$) such that $\nu(\varnothing) = 0$ and $\nu(\bigcup_1^\infty E_j) = \sum_1^\infty \nu(E_j)$ for any sequence $\{E_j\}$ of disjoint sets in $\mathcal{M}$, where the series converges absolutely if $\nu(\bigcup_1^\infty E_j)$ is finite. (We do not allow a signed measure to assume both the values $\pm\infty$ because that would lead to the ill-defined quantity $\infty - \infty$.) In discussing signed and complex measures, we will often refer to ordinary measures as *positive measures* for the sake of clarity.

The relation between signed measures and positive measures is simple:

2.20 Theorem. *If ν is a signed measure on $(X, \mathcal{M})$, there are unique positive measures ν^+ and ν^- such that $\nu = \nu^+ - \nu^-$ and $\nu^+ \perp \nu^-$.*

The decomposition $\nu = \nu^+ - \nu^-$ is called the *Jordan decomposition* of ν, and ν^+ and ν^- are called the *positive* and *negative parts* of ν. The main point of the proof of this theorem is the construction of disjoint sets $E^+, E^- \in \mathcal{M}$ such that $\nu(F) \geq 0$ whenever $F \subset E^+$, $\nu(F) \leq 0$ whenever $F \subset E^-$, and $E^+ \cup E^- = X$ (a so-called *Hahn decomposition* of X); one then defines $\nu^\pm$ by $\nu^\pm(F) = \nu(F \cap E^\pm)$.

If ν is a signed measure, we have $-\nu^-(X) \leq \nu(E) \leq \nu^+(X)$ for any measurable set E. Thus ν fails to assume the value ∞ (resp. $-\infty$) precisely when $\nu^+(X) < \infty$ (resp. $\nu^-(X) < \infty$), in which case the range of ν is actually bounded above (resp. below). Moreover, the real and imaginary parts of a complex measure are not allowed to assume infinite values, so the range of a complex measure is always a bounded subset of $\mathbb{C}$. In particular, a positive measure qualifies as a complex measure only if it is finite.

If ν is a complex measure, its real and imaginary parts ν_r and ν_i are signed measures that do not assume either value $\pm\infty$, so we have the Jordan decomposition

$$\nu = \nu_r + i\nu_i = (\nu_r^+ - \nu_r^-) + i(\nu_i^+ - \nu_i^-),$$

where $\nu_r^\pm$ and $\nu_i^\pm$ are finite positive measures.

Integration with respect to a signed or complex measure is defined in the obvious way. Namely, if ν is a signed measure, we set $L^1(\nu) = L^1(\nu^+) \cap L^1(\nu^-)$, and for $f \in L^1(\nu)$ we define $\int f\, d\nu = \int f\, d\nu^+ - \int f\, d\nu^-$. Then, if ν is a complex measure, we set $L^1(\nu) = L^1(\nu_r) \cap L^1(\nu_i)$, and for $f \in L^1(\nu)$ we define $\int f\, d\nu = \int f\, d\nu_r + i\int f\, d\nu_i$.

The notions of mutual singularity and absolute continuity also extend to signed and complex measures. Suppose μ is a positive measure on $(X, \mathcal{M})$.

If v is a signed measure on $(X, \mathcal{M})$, we say that $v \perp \mu$ (resp. $v \ll \mu$) if $v^+ \perp \mu$ and $v^- \perp \mu$ (resp. $v^+ \ll \mu$ and $v^- \ll \mu$). If v is a complex measure on $(X, \mathcal{M})$, we say that $v \perp \mu$ (resp. $v \ll \mu$) if $v_r \perp \mu$ and $v_i \perp \mu$ (resp. $v_r \ll \mu$ and $v_i \ll \mu$).

The Lebesgue-Radon-Nikodym theorem is easily extended to signed or complex measures by applying it to positive and negative parts or real and imaginary parts. To be precise, one can replace the hypothesis "v is a σ-finite measure" by either "v is a complex measure" or "v is a signed measure such that v^+ and v^- are σ-finite"; in the conclusion, the condition "$f \in L^+(X)$" is then replaced by "$f \in L^1(\mu)$" or "f is a measurable real-valued function such that at least one of $\int f^+ \, d\mu$ and $\int f^- \, d\mu$ is finite," respectively.

If v is a signed measure, the positive measure $v^+ + v^-$ is called the *total variation* of v and is denoted by $|v|$. We observe that $v \ll |v|$ and $dv/d|v| = \chi_E - \chi_F$ where $X = E \cup F$ is a Hahn decomposition of X.

The notion of total variation can be extended to complex measures, but its definition is a little more subtle. The idea is that if v is a complex measure such that $dv = f \, d\mu$ where μ is a positive measure and $f \in L^1(\mu)$, then the total variation $|v|$ should be given by $d|v| = |f| \, d\mu$. In fact, this condition specifies $|v|$ uniquely. Setting $\mu = |v_r| + |v_i|$, we clearly have $v \ll \mu$, so dv has the form $f \, d\mu$ by the Lebesgue-Radon-Nikodym theorem. Moreover, it is not hard to check that if we also have $dv = f' \, d\mu'$, then the measures $|f| \, d\mu$ and $|f'| \, d\mu'$ are equal, so the requirement that $d|v| = |f| \, d\mu$ determines $|v|$.

There are also ways of defining $|v|$ without recourse to Radon-Nikodym derivatives, but they are computationally awkward. Perhaps the most commonly found characterization of $|v|$ is this:

$$|v|(E) = \sup \left\{ \sum_1^n |v(E_j)| : n \geq 1, \ E_1, \ldots, E_n \text{ disjoint}, E = \bigcup_1^n E_j \right\}.$$

We conclude by recording two useful observations concerning total variations:

- If v is a signed or complex measure, then $L^1(v) = L^1(|v|)$, and if $f \in L^1(v)$, we have

$$\left| \int f \, dv \right| \leq \int |f| \, d|v|.$$

- If v_1 and v_2 are complex measures on $(X, \mathcal{M})$, then

$$|v_1 + v_2| \leq |v_1| + |v_2|.$$

CHAPTER 3

MEASURE AND INTEGRATION: CONSTRUCTIONS AND SPECIAL EXAMPLES

In this chapter we begin by presenting a general scheme for constructing measures. We then use it to construct Lebesgue measure and related measures on Euclidean space, and we analyze these measures and their associated integrals in some detail. We conclude with a discussion of regular Borel measures and integrals on locally compact Hausdorff spaces.

3.1 CONSTRUCTION OF MEASURES

The construction of nontrivial examples of measures is not easy. To motivate the ideas, let us consider the elementary notion of area for regions in the plane $\mathbb{R}^2$ that is defined in terms of grids of rectangles. We first define the area of a rectangle (the Cartesian product of two intervals) to be the product of the lengths of its sides; we then have a notion of area for finite unions of rectangles. Then, given a bounded set $E \subset \mathbb{R}^2$, we consider such finite unions of rectangles, $\bigcup_1^n R_j$, that approximate E from the outside ($E \subset \bigcup_1^n R_j$) or from the inside ($E \supset \bigcup_1^n R_j$). If we can find sequences of such outer and inner approximations whose areas approach the same limit, we take the area of E to be the common limit. Let us observe also that if E is contained in a rectangle R, we can pass from inner approximations to E to outer approximations to $R \setminus E$ or back by taking relative complements in R.

Here is an abstract version of the "outer approximation" procedure. We start with a space X, a family $\mathcal{E}$ of subsets of X such that $\varnothing \in \mathcal{E}$ and $X \in \mathcal{E}$, and a notion of measure for elements of $\mathcal{E}$. For present purposes, this can be any function $\rho : \mathcal{E} \to [0, \infty]$ such that $\rho(\varnothing) = 0$. We then define a preliminary notion of measure for arbitrary subsets of X in terms of outer approximations by unions of sets in $\mathcal{E}$ — and we will use countable rather than finite unions here — as follows: for any $E \subset X$, we set

$$(3.1) \qquad \mu^*(E) = \inf \left\{ \sum_1^\infty \rho(A_j) : A_j \in \mathcal{E} \text{ and } E \subset \bigcup_1^\infty A_j \right\}.$$

It is easy to verify that the function μ^* has the properties
  i. $\mu^*(\varnothing) = 0$.
  ii. If $E \subset F$, then $\mu^*(E) \leq \mu^*(F)$.
  iii. $\mu^*(\bigcup_1^\infty E_j) \leq \sum \mu^*(E_j)$ for all $E_1, E_2, \ldots \subset X$.
Any function $\mu^* : \mathcal{P}(X) \to [0, \infty]$ with these three properties is called an *outer measure*.

The best way to bring the notion of "inner measure" for a set E into play in this abstract situation is to consider the outer measure of the complement of E. That is, let μ^* be an outer measure on X, and let us suppose for the moment that $\mu^*(X) < \infty$. (In the setting of areas of sets in the plane, take X to be a large rectangle rather than the whole plane.) The "inner measure" of $E \subset X$ can then be defined as $\mu^*(X) - \mu^*(X \setminus E)$, and the condition that the inner and outer measures coincide is that $\mu^*(X) = \mu^*(E) + \mu^*(X \setminus E)$.

This observation points the way to the key concept: If μ^* is an outer measure on X, a set $E \subset X$ is called μ^*-*measurable* if

$$\mu^*(A) = \mu^*(A \cap E) + \mu^*(A \setminus E) \text{ for all } A \subset X.$$

The generalization from the case $A = X$ to an arbitrary $A \subset X$ is quite a leap, but it is justified by the following fundamental theorem, which is the basis for most constructions of measures.

3.2 Carathéodory's theorem. *If μ^* is an outer measure on X, the collection $\mathcal{M}$ of μ^*-measurable sets is a σ-algebra, and the restriction of μ^* to $\mathcal{M}$ is a complete measure.*

One general setting in which this theorem can be applied is in extending a measure from an algebra to a σ-algebra. To be precise, let $\mathcal{A}$ be an algebra of subsets of X. A function $\mu_0 : \mathcal{A} \to [0, \infty]$ will be called a *premeasure* if $\mu_0(\varnothing) = 0$ and $\mu_0(\bigcup_1^\infty A_j) = \sum_1^\infty \mu_0(A_j)$ whenever $\{A_j\}$ is a sequence of disjoint sets in $\mathcal{A}$ such that $\bigcup_1^\infty A_j \in \mathcal{A}$. Observe that premeasures are always finitely additive on $\mathcal{A}$; that is, $\mu_0(\bigcup_1^n A_j) = \sum_1^n \mu_0(A_j)$ for any disjoint $A_1, \ldots, A_n \in \mathcal{A}$, for one can take $A_j = \emptyset$ for $j > n$. Just as for measures, a premeasure μ_0 is said to be σ-*finite* if $X = \bigcup_1^\infty A_j$ where $A_j \in \mathcal{A}$ and $\mu_0(A_j) < \infty$.

3.3 Theorem. *Let $\mathcal{A}$ be an algebra of subsets of X, $\mathcal{M}$ the σ-algebra generated by $\mathcal{A}$, and μ_0 a σ-finite premeasure on $\mathcal{A}$. Then μ_0 has a unique extension to a measure μ on $\mathcal{M}$. More precisely, let μ^* be the outer measure associated to μ_0 by (3.1),*

$$\mu^*(E) = \inf \left\{ \sum_1^\infty \mu_0(A_j) : A_j \in \mathcal{A} \text{ and } E \subset \bigcup_1^\infty A_j \right\},$$

and let $\mathcal{M}^*$ be the σ-algebra of μ^*-measurable sets. Then $\mu^*|\mathcal{A} = \mu_0$ and $\mathcal{M} \subset \mathcal{M}^*$, so the extension in question is $\mu = \mu^*|\mathcal{M}$; moreover, $\mu^*|\mathcal{M}^*$ is the completion of μ.

The proofs of these theorems are straightforward but somewhat lengthy. Theorem 3.3 remains valid without the assumption of σ-finiteness, except that the extension of μ_0 to $\mathcal{M}$ may not be unique and the σ-algebra $\mathcal{M}^*$ may be larger than the μ-completion of $\mathcal{M}$.

As a first application of Theorem 3.3, we can give the construction of product measures that we described without proof in §2.4. For simplicity we restrict attention to the case of two factors. Thus, suppose $(X, \mathcal{M}, \mu)$ and $(Y, \mathcal{N}, \nu)$ are σ-finite measure spaces. Let us call any subset of $X \times Y$ of the form $A \times B$ where $A \in \mathcal{M}$ and $B \in \mathcal{N}$ a *rectangle*, and let $\mathcal{A}$ be the collection of all finite unions of disjoint rectangles. Then $\mathcal{A}$ is an algebra, and $\mathcal{M}$ is the σ-algebra generated by $\mathcal{A}$. (Finite unions of rectangles are easily expressed as finite disjoint unions, and the complement of $A \times B$ is the union of $(X \setminus A) \times B$ and $X \times (Y \setminus B)$.) Moreover, the function π defined on rectangles by $\pi(A \times B) = \mu(A)\nu(B)$ extends by additivity to a premeasure on $\mathcal{A}$. (One has to check that if a set $E \in \mathcal{A}$ is expressed as a finite disjoint union of rectangles in two different ways, $E = \bigcup_1^n R_j = \bigcup_1^m R_k'$, then $\sum_1^n \pi(R_j) = \sum_1^m \pi(R_k')$, and that π is countably additive according to the definition of a premeasure. For both purposes it is enough to show that if $A \times B = \bigcup A_j \times B_j$ where the $A_j \times B_j$ are disjoint, then $\mu(A)\nu(B) = \sum \mu(A_j)\nu(B_j)$. This follows by observing that

$$\chi_A(x)\chi_B(y) = \chi_{A \times B}(x, y) = \sum \chi_{A_j \times B_j}(x, y) = \sum \chi_{A_j}(x)\chi_{B_j}(y)$$

and integrating the functions on the far left and right with respect to x and then with respect to y, using the monotone convergence theorem to pass from finite to infinite sums.) Theorem 3.3 then guarantees that there is a unique measure on $\mathcal{M} \otimes \mathcal{N}$ that extends π.

3.2   LEBESGUE MEASURE

There are several ways of constructing Lebesgue measure on $\mathbb{R}^n$. The procedure we shall follow is first to do the case $n = 1$ and then to apply the product construction; we shall sketch three other approaches at the end of this section.

To build a theory of measure on $\mathbb{R}$ we start with intervals, and it is convenient to use half-open intervals since they nest together without overlapping. To be precise, let us call an interval $(a, b]$ that is open on the left and closed on the right — including the cases of the infinite intervals (a, ∞) and the empty interval $\varnothing$ — an *h-interval*, and let $\mathcal{A}$ be the collection of all finite unions of disjoint h-intervals. $\mathcal{A}$ is easily seen to be an algebra, and the σ-algebra that it generates is the Borel σ-algebra $\mathcal{B}_{\mathbb{R}}$.

The function that assigns to each h-interval $(a, b]$ its length $b - a$ extends to $\mathcal{A}$ by additivity, and the resulting $\lambda_0 : \mathcal{A} \to [0, \infty]$ is a premeasure. The proof that if $(a, b] = \bigcup_1^\infty (a_j, b_j]$ (disjoint union) then $b - a = \sum_1^\infty (b_j - a_j)$ is somewhat more laborious than one might expect, because the subintervals do not have to be lined up from left to right; the sequence $\{b_j\}$ may have infinitely many cluster points. Once this is done, Theorem 3.3 assures us that λ_0 extends uniquely to a Borel measure, the completion of which is *Lebesgue measure* on $\mathbb{R}$. We denote Lebesgue measure by λ and the σ-algebra of Lebesgue measurable sets by $\mathcal{L}$. Specifically, the Lebesgue measure of any $E \in \mathcal{L}$ is given by

$$(3.4) \qquad \lambda(E) = \inf \left\{ \sum_1^\infty (b_j - a_j) : E \subset \bigcup_1^\infty (a_j, b_j] \right\}.$$

This formula is often written with the h-intervals replaced by open intervals. This is possible because one can replace $(a_j, b_j]$ by $(a_j, b_j + 2^{-j} \epsilon)$ at the cost of increasing the sum of the lengths by the arbitrarily small number ϵ.

We can now form the product measure $\lambda \times \cdots \times \lambda$ on $\mathbb{R}^n$, its domain being either $\mathcal{L} \otimes \cdots \otimes \mathcal{L}$ or the smaller σ-algebra $\mathcal{B}_{\mathbb{R}^n}$. The completion of this measure, which is the same no matter which of these domains one starts with, is *Lebesgue measure* on $\mathbb{R}^n$. We denote it again by λ and its domain by $\mathcal{L}$; the relevant value of n will always be clear from the context.

We turn to a discussion of the basic properties of Lebesgue measure. It is time to introduce a definition that applies in much more general situations: Suppose X is a locally compact Hausdorff space, and $\mathcal{M}$ is a σ-algebra on X that includes the Borel algebra $\mathcal{B}_X$. A measure μ on $(X, \mathcal{M})$ is called *regular* if

i. $\mu(K) < \infty$ for every compact $K \subset X$,

ii. $\mu(E) = \inf\{\mu(U) : U \text{ open and } U \supset E\}$ for every $E \in \mathcal{M}$,

iii. $\mu(E) = \sup\{\mu(K) : K \text{ compact and } K \subset E\}$ for every $E \in \mathcal{M}$.

Properties (ii) and (iii) are called *outer regularity* and *inner regularity*, respectively.

3.5 Proposition. *Lebesgue measure on $\mathbb{R}^n$ is regular.*

Property (i) is obvious. When $n = 1$, outer regularity holds by the remarks following (3.4). For $n > 1$, Lebesgue measure on $\mathbb{R}^n$ is defined by a formula similar to (3.4) with intervals replaced by sets of the form $\prod_1^n E_j$ with $E_j \subset \mathbb{R}$, so we can use outer regularity on $\mathbb{R}$ to replace the E_j by slightly larger open sets, which gives outer regularity on $\mathbb{R}^n$. Inner regularity for bounded sets E — say, $E \subset (-c, c)^n$ — follows from outer regularity for $[-c, c]^n \setminus E$, and inner regularity for general E follows by breaking E up into countably many bounded pieces.

A consequence of regularity is that arbitrary Lebesgue measurable sets differ from Borel sets of a relatively simple form by sets of measure zero. Specifically, if $E \subset \mathcal{L}$ and $\lambda(E) < \infty$, for all $j \geq 1$ there is an open set U_j and a compact set K_j with $K_j \subset E \subset U_j$, $\lambda(U_j \setminus E) < 2^{-j}$, and $\lambda(E \setminus K_j) < 2^{-j}$. Letting $V = \bigcap_1^\infty U_j$ and $F = \bigcup_1^\infty K_j$, we obtain a G_δ set V and an F_σ set F such that $F \subset E \subset V$ and $\lambda(V \setminus E) = \lambda(E \setminus F) = 0$. It is an easy exercise to extend this result to the case $\lambda(E) = \infty$ by cutting E into bounded pieces.

The relationship between topological and measure-theoretic properties of sets is somewhat subtle, however. Given $\epsilon > 0$, let $\{x_j\}_1^\infty$ be a countable dense subset of $[0, 1]$, let I_j be the interval of length $2^{-j}\epsilon$ centered at x_j, and let $U = \bigcup_1^\infty I_j \cap (0, 1)$. Then U is open and dense in $[0, 1]$, and hence topologically "large," and $[0, 1] \setminus U$ is nowhere dense and hence topologically "small," but $\lambda(U) < \epsilon$ and $\lambda([0, 1] \setminus U) > 1 - \epsilon$.

The Lebesgue measure of any single point is obviously zero, and hence so is the Lebesgue measure of any countable set, such as the set of points in $\mathbb{R}^n$ with rational coordinates. There are also uncountable sets of measure zero. The classic example is the (standard) *Cantor set* C, which is obtained from $[0, 1]$ by removing the open middle third $(\frac{1}{3}, \frac{2}{3})$, then removing the open middle thirds $(\frac{1}{9}, \frac{2}{9})$ and $(\frac{7}{9}, \frac{8}{9})$ from the two remaining subintervals, and so on inductively. The Cantor set can be conveniently described as the set of all points in $[0, 1]$ that have a base-3 decimal expansion containing only the digits 0 and 2. It is a compact, nowhere dense set with no isolated points but no nonempty connected subsets except single points;

it has the cardinality of the continuum because the map $\sum_1^\infty b_j 3^{-j} \mapsto$ $\sum_1^\infty (b_j/2)2^{-j}$ (where $b_j = 0$ or 2) is a surjection from C onto $[0, 1]$; and it has Lebesgue measure zero because the measure of its complement $[0, 1] \setminus C$ is $1 \cdot \frac{1}{3} + 2 \cdot \frac{1}{9} + 4 \cdot \frac{1}{27} + \cdots = 1$.

The other fundamental feature of Lebesgue measure is its behavior under translations and linear transformations.

3.6 Theorem. *For $a \in \mathbb{R}^n$, define $\tau_a : \mathbb{R}^n \to \mathbb{R}^n$ by $\tau_a(x) = x + a$, and let T be an arbitrary invertible linear transformation of $\mathbb{R}^n$.*

a. *The class of Lebesgue measurable sets is invariant under τ_a and T; that is, if $E \in \mathcal{L}$ then $\tau_a(E) \in \mathcal{L}$ and $T(E) \in \mathcal{L}$.*

b. *If $E \in \mathcal{L}$, then $\lambda(\tau_a(E)) = \lambda(E)$ and $\lambda(T(E)) = |\det T|\lambda(E)$.*

c. *If f is a Lebesgue measurable function on $\mathbb{R}^n$, then so are $f \circ \tau_a$ and $f \circ T$. Moreover, if either $f \geq 0$ or $f \in L^1(\lambda)$, then $\int f \circ \tau_a \, d\lambda = \int f \, d\lambda$ and $\int f \circ T \, d\lambda = |\det T| \int f \, d\lambda$.*

The case $n = 1$ follows easily from the fact that the length of an interval is unchanged by translations and is multiplied by $|c|$ under the transformation $x \mapsto cx$, and the translation invariance for $n > 1$ then follows easily from the construction of the product measure. To analyze the behavior under linear transformations for $n > 1$, one can use the fact that every invertible matrix can be row-reduced to the identity, which implies that an invertible linear transformation is a composition of transformations of the following three types: multiplying one coordinate by a nonzero constant, adding a constant multiple of one coordinate to another coordinate, and interchanging two coordinates. The formula $\int f \circ T \, d\lambda = |\det T| \int f \, d\lambda$ for these elementary transformations reduces to the one-dimensional case by an application of Fubini's theorem, and it follows in general since the determinant of a product of transformations is the product of the determinants. The remainder of the proof consists of some technical details concerning sets of measure zero.

The translation-invariance of λ leads to an important uniqueness theorem:

3.7 Theorem. *If μ is a translation-invariant Borel measure on $\mathbb{R}^n$ such that $0 < \mu([0, 1]^n) < \infty$, then μ is a constant multiple of Lebesgue measure.*

Indeed, if $c = \mu([0, 1]^n)$, it is easy to show that $\mu(E) = c r^n = c\lambda(E)$ whenever E is a cube with rational side length r and thence that $\mu(E) = c\lambda(E)$ whenever E is a product of intervals (approximate E by unions of cubes with rational sides). It then follows from the uniqueness in Theorem 3.3 that $\mu = c\lambda$ in general.

Once one knows how Lebesgue integrals transform under linear transformations, it is easy to guess the change-of-variables formula for general differentiable transformations, as any differentiable transformation is linear at the infinitesimal level. Making the guess into a proof, however, requires some technical work.

3.8 Theorem. *Let Ω_1 and Ω_2 be open sets in $\mathbb{R}^n$. Suppose $\Phi : \Omega_1 \to \Omega_2$ is a bijection such that Φ and Φ^{-1} are both continuously differentiable, and denote the matrix $(\partial \Phi_i / \partial x_k)(x)$ of partial derivatives of Φ at x by $D_x\Phi$. If f is a Lebesgue measurable function on Ω_2, then $f \circ \Phi$ is Lebesgue measurable on Ω_1, and if either $f \geq 0$ or $f \in L^1(\Omega_2, \lambda)$, then*

$$\int_{\Omega_2} f \, d\lambda = \int_{\Omega_1} f \circ \Phi(x) |\det D_x\Phi| \, d\lambda(x).$$

The most frequently used nonlinear transformations are the polar coordinate map $(r, \theta) \to (r \cos \theta, r \sin \theta)$ in the plane and its analogues in higher dimensions (spherical coordinates in $\mathbb{R}^3$, etc.). Each of these has an integration formula associated to it according to Theorem 3.8. For many purposes, however, the most important point is the reduction of integrals of *radial* functions — that is, functions $f(x)$ that depend only on $|x|$ — to one-dimensional integrals, and this can be accomplished quite simply.

3.9 Proposition. *If f is a measurable function on $[0, \infty)$, then the function $x \mapsto f(|x|)$ is in $L^1(\mathbb{R}^n, \lambda)$ if and only if the function $r \mapsto f(r)r^{n-1}$ is in $L^1([0, \infty), \lambda)$, in which case*

$$(3.10) \qquad \int_{\mathbb{R}^n} f(|x|) \, d\lambda(x) = nc_n \int_{[0,\infty)} f(r)r^{n-1} \, d\lambda(r),$$

where c_n is the Lebesgue measure of the unit ball:
$$(3.11)$$

$$c_n = \lambda(B(1,0)) = \frac{\pi^{n/2}}{\Gamma((n/2) + 1)} = \begin{cases} \dfrac{\pi^{n/2}}{(n/2)!} & \text{if } n \text{ is even,} \\[2ex] \dfrac{2^{(n+1)/2}\pi^{(n-1)/2}}{1 \cdot 3 \cdot 5 \cdots n} & \text{if } n \text{ is odd.} \end{cases}$$

(Here $\Gamma(s) = \int_0^\infty t^{s-1}e^{-t} \, d\lambda(t)$.)

Indeed, it follows from Theorem 3.6 that the measure of a ball of radius R is $c_n R^n$, and hence, if $f = \chi_{[a,b]}$,

$$\int_{\mathbb{R}^n} f(|x|) \, d\lambda(x) = \lambda(B(b,0)) - \lambda(B(a,0)) = c_n(b^n - a^n),$$

which equals $n c_n \int_a^b r^{n-1} d\lambda(r)$ as claimed. It then follows from the uniqueness in Theorem 3.3 that (3.10) holds when f is the characteristic function of any measurable set in $[0, \infty)$, and hence in general. For the calculation of the constant $c_n = \lambda(B(1, 0))$ and a more complete discussion of integration in polar coordinates, see [6, §2.7].

Next we discuss some facts about Lebesgue integrable functions. The first one is that any function in $L^1(\lambda)$ can be approximated in the L^1 metric by continuous functions.

3.12 Proposition. *If $f \in L^1(\lambda)$ and $\epsilon > 0$, there is a continuous function g that vanishes outside a bounded set such that $\int |f - g| \, d\lambda < \epsilon$.*

By approximating f by simple functions, one reduces to the case where f is the characteristic function of a set E with $\lambda(E) < \infty$. In dimension $n = 1$, one uses the definition of λ to find a set F that is a finite disjoint union of intervals such that $\int |\chi_E - \chi_F| \, d\lambda = \lambda(E \setminus F) + \lambda(F \setminus E)$ is as small as one wishes, and then approximates χ_F by piecewise linear functions by a simple direct construction. An elaboration of this idea then gives the result for general n.

A Lebesgue measurable function f is said to be *locally integrable* if $\int_E |f| \, d\lambda < \infty$ for every bounded measurable set E. For such functions it is of interest to consider their averages over balls. That is, if f is locally integrable and $r > 0$, we define $A_r f(x)$ to be the average value of f on $B(r, x)$:

$$A_r f(x) = \frac{1}{\lambda(B(r, x))} \int_{B(r,x)} f(y) \, d\lambda(y).$$

We would expect that, for small r, $A_r f$ should be some sort of smoothed-out approximation to f. Indeed, it is not hard to see that $A_r f(x)$ is jointly continuous in $r \in (0, \infty)$ and $x \in \mathbb{R}^n$, and if f is itself continuous, then $A_r f(x) \to f(x)$ for all x as $r \to 0$. It is a remarkable fact that $A_r f(x) \to f(x)$ as $r \to 0$ at least for almost every x when f is any locally integrable function.

In fact, we can say something stronger. The condition $A_r f(x) \to f(x)$ can be restated as

$$\lim_{r \to 0} \frac{1}{\lambda(B(r, x))} \int_{B(r,x)} [f(y) - f(x)] \, d\lambda(y) = 0.$$

The first strengthening is that we can replace $f(y) - f(x)$ by its absolute value, and the second is that we can replace the balls $B(r, x)$ by other sets that approach $\{x\}$ in a suitable way as $r \to 0$. The precise definition is this:

A family $\{E_r\}_{r>0}$ of measurable sets is said to *shrink nicely to* x as $r \to 0$ if $E_r \subset B(r, x)$ and there is a constant $\alpha > 0$ such that $\lambda(E_r) \geq \alpha\lambda(B(r, x))$ for all $r > 0$.

3.13 The Lebesgue differentiation theorem. *Let f be a locally integrable function on $\mathbb{R}^n$, and define the* Lebesgue set *of f to be*

$$L_f = \left\{ x : \lim_{r \to 0} \frac{1}{\lambda(B(r, x))} \int_{B(r,x)} |f(y) - f(x)|\, d\lambda(y) = 0 \right\}.$$

Then $\lambda(\mathbb{R}^n \setminus L_f) = 0$, and for every $x \in L_f$ we have

$$\lim_{r \to 0} \frac{1}{\lambda(E_r)} \int_{E_r} |f(y) - f(x)|\, d\lambda(y) = 0$$

and

$$\lim_{r \to 0} \frac{1}{\lambda(E_r)} \int_{E_r} f(y)\, d\lambda(y) = f(x)$$

for every family of sets $\{E_r\}$ that shrinks nicely to x.

As we observed in the preceding chapter, for the purposes of integration theory one can modify functions on sets of measure zero without affecting anything, so it might seem that it makes little sense to speak of the pointwise values of an element of L^1, or more generally of a locally integrable function. But this theorem shows that if $[f]$ is an equivalence class of locally integrable functions (the equivalence being almost-everywhere equality), there is a canonical representative $f_0 \in [f]$ defined almost everywhere; namely, one picks any $f \in [f]$ and sets $f_0(x) = \lim_{r \to 0} A_r f(x)$ for all x such that the limit exists. This limit is independent of the choice of $f \in [f]$; it actually equals $f(x)$ for any x at which f is continuous.

The proof of the Lebesgue differentiation theorem uses a technical tool that is of interest in its own right. For $\phi \in L^1(\lambda)$, one defines the *Hardy-Littlewood maximal function* $H\phi$ by

$$H\phi(x) = \sup_{r>0} A_r|\phi|(x) = \sup_{r>0} \frac{1}{\lambda(B(r, x))} \int_{B(r,x)} |\phi(y)|\, d\lambda(y)$$

and proves that there is a constant $c > 0$ such that for all $\phi \in L^1(\lambda)$ and all $\alpha > 0$ we have

$$(3.14) \qquad \lambda\big(\{x : H\phi(x) > \alpha\}\big) \leq \frac{c}{\alpha} \int |\phi(x)|\, d\lambda(x).$$

(The Hardy-Littlewood maximal function and the estimate (3.14) are proto-
types of a large family of maximal functions and estimates that have proved
to be of great utility in modern analysis.)

Now, to prove the Lebesgue differentiation theorem, it is enough to assume that $f \in L^1(\lambda)$, in which case one can find a continuous g such
that $\int |f - g| \, d\lambda$ is as small as one wishes, by Proposition 3.12. Since
$A_r g(x) \to g(x)$ for all x, one can conclude that the measure of the set

$$\left\{ x : \limsup_{r \to 0} |A_r f(x) - f(x)| > \alpha \right\}$$

is arbitrarily small for any $\alpha > 0$ by applying the estimate (3.14) with
$\phi = f - g$, and it follows easily that $A_r f(x) \to f(x)$ almost everywhere.
Once this is known, a clever little argument yields the stronger assertion
that $\lambda(\mathbb{R}^n \setminus L_f) = 0$; the replacement of $B(r, x)$ by the more general sets
E_r is then easy.

We can use the Lebesgue differentiation theorem to give a pointwise
formula for Radon-Nikodym derivatives of regular Borel measures on $\mathbb{R}^n$.
In a little more generality, we have the following:

3.15 Theorem. *Let v be a signed or complex Borel measure on $\mathbb{R}^n$ such
that $|v|$ is regular, and let $v = v_s + v_{ac}$ be its Lebesgue decomposition with
respect to Lebesgue measure λ. Then for λ-almost every $x \in \mathbb{R}^n$ we have*

$$\frac{dv_{ac}}{d\lambda}(x) = \lim_{r \to 0} \frac{v(E_r)}{\lambda(E_r)}$$

for every family $\{E_r\}_{r>0}$ that shrinks nicely to x.

To prove this, one verifies the easily believable fact that for almost every
x, $v_s(E_r)/\lambda(E_r) \to 0$ as $r \to 0$ when $\{E_r\}$ shrinks nicely to x. The
finiteness of $|v|$ on compact sets implies that $dv_{ac}/d\lambda$ is locally integrable,
and the result follows by applying the Lebesgue differentiation theorem.

We conclude this section by sketching some alternative constructions
of Lebesgue measure and Lebesgue integrals. First, it is possible to define
Lebesgue measure on $\mathbb{R}^n$ directly without going through the special case
$n = 1$ and using the product construction. Let $\mathcal{E}$ be the collection of subsets
of $\mathbb{R}^n$ that are finite disjoint unions of sets of the form $\prod_1^n I_j$ where each
I_j is a bounded interval. One builds up the definition of λ in a sequence of
steps by the following prescriptions (which are theorems in our treatment
of λ):

- $\lambda(\prod_1^n I_j) = \prod_1^n l(I_j)$ where l denotes length, and λ is then extended
  to sets in $\mathcal{E}$ by additivity.

- If $U \subset \mathbb{R}^n$ is open, $\lambda(U) = \sup\{\lambda(E) : E \in \mathcal{E} \text{ and } E \subset U\}$.

- If $K \subset \mathbb{R}^n$ is compact, $\lambda(K) = \inf\{\lambda(E) : E \in \mathcal{E} \text{ and } E \supset K\}$.

- A bounded set $B \subset \mathbb{R}^n$ is Lebesgue measurable if and only if

  $$\inf\{\lambda(U) : U \text{ open and } U \supset B\} = \sup\{\lambda(K) : K \text{ compact and } K \subset B\},$$

  in which case $\lambda(B)$ is this common value.

- Finally, an arbitrary $A \subset \mathbb{R}^n$ is Lebesgue measurable if and only if $A \cap [-c, c]^n$ is measurable for all $c > 0$, in which case

  $$\lambda(A) = \lim_{c \to \infty} \lambda(A \cap [-c, c]^n).$$

For the details of this approach, see Jones [9] or Fleming [5].

Another approach is first to define the Riemann integral as in advanced calculus for continuous functions on $\mathbb{R}^n$ that vanish outside a bounded set, obtaining thereby a positive linear functional on the space of all such functions, and then to invoke the Riesz representation theorem that we shall present in §3.6 to produce Lebesgue measure.

Finally, there is a clever way, due to Henstock and Kurzweil, of modifying the definition of the Riemann integral so as to yield an integral that is even more general than the Lebesgue integral without doing any measure theory first. (One then recovers the measure of a set E as the integral of χ_E.) Here is how it works; for simplicity we restrict attention to functions on a bounded interval $[a, b]$.

A *tagged partition* of $[a, b]$ is a finite sequence $\{x_j\}_0^N$ such that $a = x_0 < x_1 < \cdots < x_N = b$ together with another finite sequence $\{t_j\}_1^N$ such that $t_j \in [x_{j-1}, x_j]$. A *gauge* on $[a, b]$ is an arbitrary function $\delta : [a, b] \to (0, \infty)$. If P is a tagged partition and δ is a gauge, P is called δ-*fine* if $x_j - x_{j-1} < \delta(t_j)$ for all j. Now, if f is a real-valued function on $[a, b]$, every tagged partition P of $[a, b]$ defines a Riemann sum for f, namely, $\Sigma_P f = \sum_1^n f(t_j)(x_j - x_{j-1})$. We say that f is *Henstock-Kurzweil integrable* if there exists $c \in \mathbb{R}$ with the property that for every $\epsilon > 0$ there is a gauge δ_ϵ such that for any δ_ϵ-fine partition P of $[a, b]$ we have $|\Sigma_P f - c| < \epsilon$, in which case c is called the *Henstock-Kurzweil integral* of f.

If one allows only constant gauges, the result is the Riemann integral. The simple change of passing to arbitrary gauges, however, has the effect that every Lebesgue integrable function on $[a, b]$ is also Henstock-Kurzweil integrable, and the two integrals coincide. Moreover, the class of Henstock-Kurzweil integrable functions properly includes $L^1([a, b], \lambda)$. It is still contained in the class of Lebesgue measurable functions, and its intersection with the class of nonnegative functions is still $L^1 \cap L^+$, but it includes some functions f such that $\int f^+ \, d\lambda = \int f^- \, d\lambda = \infty$, whose integral is only "conditionally convergent."

This combination of simplicity and generality has won the Henstock-Kurzweil integral a fan club that advocates presenting it instead of the Lebesgue integral in real analysis courses. However, although the Henstock-Kurzweil integral can be adapted to functions on $\mathbb{R}^n$ without much trouble, it does not generalize to other spaces in a clean way; and its conditionally convergent integrals, apart from ones that can be obtained from Lebesgue integrals by a simple limiting procedure as in the "improper integrals" of calculus, are useful only in a handful of situations. (We shall mention one in the next section.) For these reasons the present author is not a member of the fan club. Readers who want to learn more, however, may consult Bartle [1] for an introductory account and McLeod [13] for a more extensive treatment.

3.3 REGULAR BOREL MEASURES AND FUNCTIONS ON THE REAL LINE

If μ is a regular Borel measure on $\mathbb{R}$, the function $F : \mathbb{R} \to \mathbb{R}$ defined by

$$F(x) = \begin{cases} -\mu((x, 0]) & \text{if } x < 0, \\ 0 & \text{if } x = 0, \\ \mu((0, x]) & \text{if } x > 0 \end{cases}$$

is increasing and right continuous (i.e., $F(x) \to F(a)$ as x approaches a from the right). It is called the *cumulative distribution function* of μ (with base point 0); we have $\mu((a, b]) = F(b) - F(a)$ for all $a < b$.

Conversely, if F is any increasing, right continuous function on $\mathbb{R}$, there is a unique Borel measure μ on $\mathbb{R}$ such that $\mu((a, b]) = F(b) - F(a)$ for all $a < b$, and this measure is regular. The construction of μ is entirely parallel to the construction of Lebesgue measure on $\mathbb{R}$ — the special case $F(x) = x$ — using Theorem 3.3 as in §3.2. Outer regularity holds by the remarks following (3.4), which still apply in this more general situation since F is right continuous, and inner regularity follows as for Lebesgue measure. (Incidentally, this shows that any Borel measure μ on $\mathbb{R}$ that is finite on compact sets is automatically regular: Given such a μ, one can define F as above and then recover μ from F by the analogue of (3.4), from which regularity follows.)

If μ and F are related in this way, the integral $\int f \, d\mu$ is classically denoted by $\int f \, dF$ or $\int f(x) \, dF(x)$. (The intuition is that the differential dx in the ordinary integral is replaced by $dF(x) = F(x + dx) - F(x)$.) These integrals are known as *Lebesgue-Stieltjes integrals*.

Suppose F is an increasing function on $\mathbb{R}$. Then the left- and right-hand limits

$$(3.16) \qquad F(x-) = \lim_{\epsilon \searrow 0} F(x - \epsilon), \qquad F(x+) = \lim_{\epsilon \searrow 0} F(x + \epsilon)$$

exist at every x, and we have $F(x) = F(x-) = F(x+)$ except perhaps at countably many x. (In any bounded interval $[-c, c]$, for each $n \geq 1$ there can only be finitely many x at which the jump $F(x+) - F(x-)$ is greater than $1/n$, as the sum of all the jumps in this interval is at most $F(c) - F(-c)$. Thus the set of points at which F is discontinuous is at most countable.) But more is true. In the following theorem and elsewhere in this section, "almost everywhere" means "almost everywhere with respect to Lebesgue measure" unless otherwise specified.

3.17 Proposition. *If $F : \mathbb{R} \to \mathbb{R}$ is increasing, then F is differentiable almost everywhere.*

To prove this, let $G(x) = F(x+)$, which is right continuous, and let μ be the measure such that $\mu((a, b]) = G(b) - G(a)$. Since $G(x + h) - G(x)$ equals $\mu((x, x + h])$ if $h > 0$ and $-\mu((x + h, x])$ if $h < 0$, and the sets $(x, x+h]$ or $(x+h, x]$ shrink nicely to x as $h \to 0$, it follows from Theorem 3.15 that G' exists almost everywhere. Moreover, $G(x) = F(x)$ except at a countable set of points $x_1, x_2, \ldots$, and we can see that $(G - F)'$ exists and equals zero almost everywhere by applying Theorem 3.15 to the measure $\sum [G(x_j) - F(x_j)] \delta_{x_j}$, where δ_x is the point mass at x.

Now suppose ν is a complex measure on $\mathbb{R}$. Since ν does not assume infinite values, we can use $-\infty$ rather than 0 as a base point and define the cumulative distribution function of ν by

$$F(x) = \nu((-\infty, x]).$$

Then F is a bounded, right continuous complex-valued function on $\mathbb{R}$, and the Jordan decomposition $\nu = (\nu_r^+ - \nu_r^-) + i(\nu_i^+ - \nu_i^-)$ leads to a corresponding decomposition $F = (F_r^+ - F_r^-) + i(F_i^+ - F_i^-)$ where $F_r^\pm$ and $F_i^\pm$ are increasing. However, the class of functions that can be represented in this way has another characterization that is of interest in its own right.

The *total variation* of a function $F : [a, b] \to \mathbb{C}$ on the interval $[a, b]$ is the quantity $T_F|_a^b$ defined as

$$\sup \left\{ \sum_1^n |F(x_j) - F(x_{j-1})| : n \geq 1, \, a = x_0 < x_1 < \cdots < x_n = b \right\},$$

and F is said to be *of bounded variation on* $[a, b]$ if $T_F|_a^b < \infty$. For example, if F is real-valued and has only a finite number of local maxima and minima on $[a, b]$, then F is of bounded variation on $[a, b]$, and the supremum defining $T_F|_a^b$ is achieved for any partition $\{x_j\}_0^n$ that includes these local extrema. On the other hand, it is an easy exercise to show that the function $F(x) = x \sin(1/x)$ ($F(0) = 0$) is continuous but not of bounded variation on any interval containing 0.

Let us consider a slight modification of this idea. If $F : \mathbb{R} \to \mathbb{C}$, the *total variation function* of F is the increasing function $T_F : \mathbb{R} \to [0, \infty]$ defined by

$$T_F(x) = \sup \left\{ \sum_1^n |F(x_j) - F(x_{j-1})| : n \geq 1, \, x_0 < \cdots < x_n = x \right\}.$$

We say that F is *of bounded variation* (on $\mathbb{R}$) if $T_F(x) < \infty$ for all x and $\lim_{x \to \infty} T_F(x) < \infty$, and we denote the set of all functions of bounded variation by BV. If $F \in BV$, then F is of bounded variation on every finite interval, and $T_F|_a^b = T_F(b) - T_F(a)$; on the other hand, if F is of bounded variation on $[a, b]$ and we extend F to $\mathbb{R}$ by the prescription $F(x) = F(a)$ for $x < a$ and $F(x) = F(b)$ for $x > b$, then $F \in BV$.

It is not hard to verify that if $F \in BV$ is real-valued, then the functions $T_F + F$ and $T_F - F$ are increasing, so that F is the difference of the two increasing functions $\frac{1}{2}(T_F + F)$ and $\frac{1}{2}(T_F - F)$. This representation of F is called the *Jordan decomposition* of F. (This is the decomposition actually introduced by Jordan; the measure-theoretic analogue introduced in §2.5 came later.) Moreover, a complex-valued F belongs to BV if and only if its real and imaginary parts do. With these facts in mind, the preceding results about increasing functions immediately yield analogous facts about functions in BV:

3.18 Proposition. *If $F \in BV$, then the left- and right-hand limits $F(x-)$ and $F(x+)$ exist for every x, as do the limits*

$$F(\pm\infty) = \lim_{x \to \pm\infty} F(x).$$

Moreover, the set of points at which F is discontinuous is at most countable, and $F'(x)$ exists for almost every x.

We can now complete this circle of ideas by passing from functions of bounded variation back to Borel measures. We need a slight modification of the space BV: the space of *normalized* functions of bounded variation,

$$NBV = \{F \in BV : F \text{ is right continuous and } F(-\infty) = 0\}.$$

The spaces BV and NBV are not very different. If F is any function in BV, the function $G(x) = F(x+) - F(-\infty)$ is in NBV; moreover, $G(x) = F(x) - F(-\infty)$ except perhaps at countably many x, and $G' = F'$ almost everywhere.

3.19 Theorem. *If ν is a complex Borel measure on $\mathbb{R}$ and $F(x) = \nu((-\infty, x])$, then $F \in NBV$. Conversely, if $F \in NBV$, there is a unique complex Borel measure ν_F such that $F(x) = \nu_F((-\infty, x])$; moreover, $|\nu_F| = \nu_{T_F}$.*

Here $|\nu_F|$ is the total variation of ν_F as defined in §2.5. The first two assertions of this theorem are almost immediate from what we have said already; only the proof that $|\nu_F| = \nu_{T_F}$ requires some work.

Finally, we take a closer look at absolute continuity. Like the notion of Jordan decomposition, absolute continuity can be defined either measure-theoretically or as a purely real-variable concept. For the latter, the definition is as follows: A function $F : \mathbb{R} \to \mathbb{C}$ is called *absolutely continuous* if for every $\epsilon > 0$ there is a $\delta > 0$ such that for any finite set of disjoint intervals $(a_1, b_1), \ldots, (a_n, b_n)$,

$$\sum_1^n (b_j - a_j) < \delta \implies \sum_1^n |F(b_j) - F(a_j)| < \epsilon.$$

More generally, F is said to be absolutely continuous on an interval $[a, b]$ if this condition is satisfied with the restriction that $(a_j, b_j) \subset [a, b]$ for all j. Every absolutely continuous function is uniformly continuous, and if F has a bounded everywhere-defined derivative then F is absolutely continuous (by the mean value theorem).

3.20 Theorem. *Suppose $F \in NBV$, and let ν_F be the complex Borel measure such that $\nu_F((a, b]) = F(b) - F(a)$ for $a < b$. Then:*
a. *$F' \in L^1(\lambda)$.*
b. *$\nu_F \perp \lambda$ if and only if $F' = 0$ almost everywhere.*
c. *$\nu_F \ll \lambda$ if and only if F is absolutely continuous if and only if $F(x) = \int_{-\infty}^x F'(t)\, d\lambda(t)$ for all x.*

The proof is largely a matter of applying Theorem 3.15 to ν_F. An easy modification of this result to deal with functions on bounded intervals rather than on the whole line yields the following **fundamental theorem of calculus for Lebesgue integrals**:

3.21 Theorem. *If $F : [a, b] \to \mathbb{C}$ ($-\infty < a < b < \infty$), the following are equivalent:*

  a. *F is absolutely continuous on $[a, b]$.*
  b. *$F(x) - F(a) = \int_a^x f(t)\,dt$ for some $f \in L^1([a, b], \lambda)$.*
  c. *F is differentiable almost everywhere on $[a, b]$, $F' \in L^1([a, b], \lambda)$, and*
     $$F(x) - F(a) = \int_a^x F'(t)\,d\lambda(t).$$

It must be emphasized that if F is merely known to be continuous and almost everywhere differentiable, it does not follow that $F(x) - F(a) = \int_a^x F'(t)\,d\lambda(t)$. Recall the Cantor set C and the map

$$\sum_1^\infty b_j 3^{-j} \mapsto \sum_1^\infty \frac{b_j}{2} 2^{-j} \qquad (b_j = 0 \text{ or } 2)$$

that maps C onto $[0, 1]$. This map assigns equal values to the endpoints of each of the open "middle third" intervals missing from C, so it can be extended to a function $F : [0, 1] \to [0, 1]$ that is constant on each of these intervals. This F is increasing and continuous, and its derivative exists and is zero almost everywhere on $[0, 1]$ (in fact, on $[0, 1] \setminus C$); but $F(1) - F(0) = 1$. The corresponding measure μ_F is an example of a measure that is singular with respect to Lebesgue measure but has no discrete part (i.e., $\mu_F(\{x\}) = 0$ for every x).

What can we say if we assume simply that F is everywhere differentiable? The trouble is that F' can be quite wild. It is true but not at all obvious that if $F' \in L^1([a, b], \lambda)$, then F is absolutely continuous on $[a, b]$ and $F(b) - F(a) = \int_a^b F'(t)\,d\lambda(t)$; see Rudin [17, Theorem 7.21] for a proof. But F' need not even be in L^1. For example, if $F(x) = x^2 \sin(x^{-2})$ ($F(0) = 0$), then $F'(0) = 0$ but F' is not in L^1 on any interval that contains 0, and there are other examples with more complicated singularities. Here the Henstock-Kurzweil integral comes to the rescue: if F is everywhere differentiable on $[a, b]$, then F' is Henstock-Kurzweil integrable on $[a, b]$ and its Henstock-Kurzweil integral over $[a, b]$ equals $F(b) - F(a)$.

3.4 HAUSDORFF MEASURE

Analysis on n-dimensional Euclidean space often involves the consideration of sets of lower dimension, such as curves and surfaces in $\mathbb{R}^3$ and their analogues in other dimensions, known in general as *submanifolds*. Given a smooth k-dimensional submanifold S of $\mathbb{R}^n$, classical methods from calculus and geometry yield a natural geometric notion of k-dimensional measure for subsets of S and hence a theory of integration for functions on S. However, there is also a measure-theoretic approach due to Hausdorff that gives the k-dimensional measure for all k-dimensional submanifolds at once, and also comprises a generalization that leads to the interesting notion of sets of fractional dimension.

The intuition behind Hausdorff measure is that for k-dimensional sets of a particular shape, such as balls or cubes, the k-dimensional measure of the set is proportional to the kth power of its diameter

$$\operatorname{diam} E = \sup\{|x - y| : x, y \in E\}.$$

If S is a smooth k-dimensional submanifold of $\mathbb{R}^n$ and $x \in S$, the intersection of a small ball $B(r, x)$ ($r \ll 1$) with S is nearly a k-dimensional ball. Hence, if one covers S in a reasonably efficient way by such balls, the sum of the kth powers of their diameters should yield an approximation to the k-dimensional measure of S, at least up to a proportionality constant. Moreover, as long as one is using the diameter of a set as a measure of its size, the precise shape of the set is irrelevant, so there is no real reason to use only balls for such coverings.

With these preliminary remarks in mind, here are the definitions. Let p be an arbitrary real number between 0 and n. For $E \subset \mathbb{R}^n$ and $\delta > 0$, we set

$$(3.22) \quad H_{p,\delta}(E) = \inf \left\{ \sum_1^\infty (\operatorname{diam} B_j)^p : E \subset \bigcup_1^\infty B_j \text{ and } \operatorname{diam} B_j \le \delta \right\}.$$

Here the B_j's are arbitrary subsets of $\mathbb{R}^n$, but they could be assumed to be closed (since the diameter of a set is the diameter of its closure) or open (since arbitrary sets can be embedded in open sets of slightly larger diameter). $H_{p,\delta}$ is an outer measure on $\mathbb{R}^n$, as (3.22) is an instance of the general paradigm (3.1) for creating outer measures. As δ decreases, $H_{p,\delta}(E)$ increases because the infimum is being taken over a smaller family of coverings, so the limit

$$H_p(E) = \lim_{\delta \to 0} H_{p,\delta}(E)$$

exists (it may be ∞). $H_p(E)$ is the *p-dimensional Hausdorff outer measure* of E. H_p is again an outer measure, and one can show that every Borel set is H_p-measurable. Hence, the restriction of H_p to $\mathcal{B}_{\mathbb{R}^n}$ is a measure, *p-dimensional Hausdorff measure*, that we still denote by H_p.

Since the diameter of a set $B \subset \mathbb{R}^n$ is invariant under any transformation of $\mathbb{R}^n$ that preserves distances, the same is true of H_p; thus H_p has the geometrically desirable property of being invariant under translations, rotations, and reflections. In particular, the translation-invariance for $p = n$ implies that H_n is proportional to Lebesgue measure on $\mathbb{R}^n$:

$$H_n = a_n \lambda.$$

To see this, one has merely to verify that $0 < H_n([0, 1]^n) < \infty$ and invoke Theorem 3.7. The proportionality constant turns out to be the reciprocal of the volume of a ball of diameter 1:

$$a_n = \frac{1}{\lambda(B(\frac{1}{2}, 0))} = \frac{2^n}{c_n},$$

where c_n is given by (3.11). (See Falconer [3, §1.4] for the proof.) For this reason, some people incorporate a factor of $1/a_p = \pi^{p/2}/2^p \Gamma((p/2) + 1)$ into the definition of H_p.

It now follows easily that if k is an integer with $0 < k < n$ and V is a k-dimensional vector subspace of $\mathbb{R}^n$, the restriction of H_k to Borel subsets of V coincides with $c_k \lambda_V$, where λ_V is the Lebesgue measure on V obtained by identifying V with $\mathbb{R}^k$ by choosing an orthonormal basis for V. More generally, if M is a (piece of a) smooth k-dimensional submanifold of $\mathbb{R}^n$ parametrized by a smooth one-to-one map from an open set $U \subset \mathbb{R}^k$ into $\mathbb{R}^n$, there is a change-of-variable formula that relates integrals over M to integrals over U (both with respect to H_k). The upshot is that the restriction of H_k to Borel subsets of M coincides with the surface measure on M given by Riemannian geometry, up to the normalization factor c_k.

However, Hausdorff measure H_p is also of interest when p is not an integer. It is easy to verify that for any $E \in \mathcal{B}_{\mathbb{R}^n}$ we have

$$\sup\{p : H_p(E) = \infty\} = \inf\{p : H_p(E) = 0\}.$$

This number is called the *Hausdorff dimension* of E, which we denote by $\dim_H(E)$. (When $p = \dim_H(E)$, $H_p(E)$ can be any number in $[0, \infty]$.)

Sets of Hausdorff dimension p can be constructed for any real $p \in [0, n]$ by considering sets with suitable self-similarity properties. Here is a construction that yields many interesting examples of such sets. Given a number $r \in (0, 1)$, let us define a *similitude with scaling factor r* to be a map $S : \mathbb{R}^n \to \mathbb{R}^n$ that is the composition of a rigid motion (i.e., the composition of a translation, a rotation, and possibly a reflection) with dilation by r (i.e., the map $x \mapsto rx$). If $\mathbf{S} = (S_1, \ldots, S_k)$ is a family of similitudes with scaling factor r and $E \subset \mathbb{R}^n$, let

$$S(E) = S^1(E) = \bigcup_1^k S_j(E), \qquad S^m(E) = S(S^{m-1}(E)) \text{ for } m > 1.$$

3.23 Theorem. *Let* $S = (S_1, \ldots, S_k)$ *be a family of similitudes with scaling factor* r. *Suppose there is an open set* U *such that*

$$S(U) \subset U \quad and \quad S_i(U) \cap S_j(U) = \varnothing \text{ for } i \neq j,$$

and let $\overline{U}$ *be the closure of* U. *Then* $F = \bigcap_{m=1}^\infty S^m(\overline{U})$ *is a nonempty compact set such that* $S(F) = F$, *and* $\dim_H(F) = \log_{1/r} k$.

The condition that $S(F) = F$ means that F is the union of k copies of itself scaled down by a factor of r, and the condition that $S_i(U) \cap S_j(U) = \varnothing$ guarantees that they have negligibly small overlap; this is what we mean by saying that F is "self-similar."

For example, if we take $n = 1$, $k = 2$, $r = \frac{1}{3}$, $S_1(x) = \frac{1}{3}x$, $S_2(x) = \frac{1}{3}(x + 2)$ and $U = (0, 1)$, the resulting set F is the standard Cantor set C, which therefore has Hausdorff dimension $\log_3 2$. Another well-known example is the Sierpiński gasket G, obtained by starting with a solid triangle in the plane, dividing it into four congruent subtriangles by bisecting the sides, deleting the middle triangle, and iterating; see Figure 3.1. Here $n = 2$, $k = 3$, $r = \frac{1}{2}$, and U is the interior of the original triangle; it follows that $\dim_H(G) = \log_2 3$. (We leave it to the reader to write out the three similitudes S_1, S_2, S_3.)

FIGURE 3.1. The first three approximations to the Sierpiński gasket.

3.5 REGULAR BOREL MEASURES
ON LCH SPACES

Let X be a locally compact Hausdorff (LCH) space, and let $C_c(X)$ be the space of continuous functions on X with compact support. (The existence of many such functions is guaranteed by Theorem 1.15.) If μ is a Borel measure on X such that $\mu(K) < \infty$ for every compact $K \subset X$, then

$C_c(X) \subset L^1(\mu)$, and the formula $I(f) = \int f \, d\mu$ defines a linear functional on $C_c(X)$ that is *positive* in the sense that $I(f) \geq 0$ whenever $f \geq 0$. It is a fundamental fact, and a rich source of measures, that every positive linear functional on $C_c(X)$ is of this form, and that the measure μ can be taken to be regular. (We recall from §3.2 that a regular Borel measure is one that is finite on compact sets and has the property that arbitrary Borel sets can be approximated in measure from the outside by open sets and from the inside by compact sets.)

Here is the precise theorem. We employ the notation $f \prec U$ to mean that $0 \leq f \leq 1$ and $\text{supp}(f) \subset U$, and we say that a topological space is *σ-compact* if it is a countable union of compact sets.

3.24 The Riesz representation theorem. *Suppose that X is a σ-compact LCH space and I is a linear functional on $C_c(X)$ such that $I(f) \geq 0$ whenever $f \geq 0$. Then there is a unique regular Borel measure μ on X such that $I(f) = \int f \, d\mu$ for all $f \in C_c(X)$. Moreover, μ satisfies*

$$(3.25) \quad \mu(U) = \sup\{I(f) : f \in C_c(X), \ f \prec U\} \textit{ for all open } U \subset X$$

and

$$(3.26) \quad \mu(K) = \inf\{I(f) : f \in C_c(X), \ f \geq \chi_K\} \textit{ for all compact } K \subset X.$$

To call this result the Riesz representation theorem is a historical oversimplification, as Riesz did only the case $X = [0, 1]$, and several other people were involved in generalizing the theorem to its present form. It remains true without the σ-compactness assumption, except that in general the assertion that μ is regular must be weakened when μ is not σ-finite.

To prove existence, one begins by taking (3.25) as a definition of $\mu(U)$ when U is open and then defining

$$\mu^*(E) = \inf\{\mu(E) : U \supset E, \ U \text{ open}\}$$

for arbitrary $E \subset X$, observing that $\mu^*(U) = \mu(U)$ if U is open. One verifies that μ^* is an outer measure and that every open set is μ^*-measurable. It then follows from Carathéodory's theorem that the restriction of μ^* to $\mathcal{B}_X$ (which we denote by μ) is a Borel measure, which is outer regular by definition. One then verifies (3.26), which implies that μ is finite on compact sets, and inner regularity follows from outer regularity as in the proof of Proposition 3.5. (This is where σ-compactness is needed.) Finally, one shows that $I(f) = \int f \, d\mu$ for $f \in C_c(X)$.

Fleshing out this outline into a real proof involves quite a bit of work. Uniqueness is much easier: Suppose μ is a regular Borel measure with

$I(f) = \int f \, d\mu$ for $f \in C_c(X)$. If U is open and $K \subset U$ is compact, by Theorem 1.15 there exists $f \in C_c(X)$ with $f \prec U$ and $f = 1$ on K, so that $\mu(K) \leq I(f) \leq \mu(U)$. It follows from inner regularity that (3.25) holds, so μ is determined by I on open sets, and hence on all Borel sets by outer regularity.

The Riesz representation theorem provides an alternative construction of the product of regular Borel measures. In brief, suppose μ and ν are regular Borel measures on the σ-compact LCH spaces X and Y. One verifies that for $f \in C_c(X \times Y)$ the iterated integrals $\iint f \, d\mu \, d\nu$ and $\iint f \, d\nu \, d\mu$ are equal; they define a positive linear functional on $C_c(X \times Y)$ and hence a regular Borel measure $\mu \widehat{\times} \nu$ on $X \times Y$. The same idea works for products of more than two spaces, and even for products of infinitely many spaces X_α when they are all compact and the measures μ_α on them satisfy $\mu_\alpha(X_\alpha) = 1$. The product measure obtained by this procedure is defined on the Borel σ-algebra on the product space, which in general is strictly larger than the product of the Borel σ-algebras on the factors, and it agrees on the latter with the product measure defined in §2.4. This enlargement of the domain is significant in some of the applications to probability theory.

Regularity of a Borel measure has much the same consequences in general that we recorded for Lebesgue measure in §3.2. In particular, there is an analogue of Proposition 3.12:

3.27 Proposition. *Suppose μ is a regular Borel measure on the LCH space X, or the completion of such a measure. If $f \in L^1(\mu)$ and $\epsilon > 0$, there exists $\phi \in C_c(X)$ such that $\int |\phi - f| \, d\mu < \epsilon$.*

By approximating f with simple functions we can reduce to the case $f = \chi_E$ where $\mu(E) < \infty$. By regularity, there are a compact set K and an open set U such that $K \subset E \subset U$ and $\mu(U \setminus K) < \epsilon$, and by Theorem 1.15 there exists $g \in C_c(X)$ with $g \prec U$ and $g = 1$ on K; it follows immediately that $\int |g - \chi_E| \, d\mu < \epsilon$.

There is also a deeper and stronger theorem about approximation by continuous functions:

3.28 Lusin's theorem. *Suppose μ is a regular Borel measure on the LCH space X, or the completion of such a measure, and $f : X \to \mathbb{C}$ is a μ-measurable function that vanishes outside a set E with $\mu(E) < \infty$. Then for any $\epsilon > 0$ there exists $\phi \in C_c(X)$ such that $\phi = f$ except on a set of measure at most ϵ. If $|f(x)| \leq C$ for all x, then ϕ can be taken to satisfy $|\phi(x)| \leq C$ for all x too.*

One final remark: In §3.3 we observed that every Borel measure μ on $\mathbb{R}$ that is finite on compact sets is automatically regular. In fact, the same is true on any LCH space X in which every open set is σ-compact. (This condition is stronger than the mere σ-compactness of X itself.) Proving this directly is not easy; the simplest way involves an application of the Riesz representation theorem to the functional $I(f) = \int f \, d\mu$.

CHAPTER 4

RUDIMENTS OF FUNCTIONAL ANALYSIS

Functional analysis — the meeting ground of analysis and linear algebra, mostly in an infinite-dimensional setting — is a vast subject, and this brief account does no more than scratch the surface. Our object is simply to introduce some basic concepts that are of wide utility and a few fundamental theorems: just enough to support the material in the last two chapters of this book. For those who want to learn more there are many books available; Reed and Simon [14] and Rudin [18] are among the best.

4.1 NORMED VECTOR SPACES AND BOUNDED LINEAR MAPS

Let $\mathcal{X}$ be a vector space over the field $\mathbb{F}$, where $\mathbb{F}$ is either $\mathbb{R}$ or $\mathbb{C}$. A *seminorm* on $\mathcal{X}$ is a function $p : \mathcal{X} \to [0, \infty)$ such that

  i. $p(x + y) \leq p(x) + p(y)$ for all $x, y \in \mathcal{X}$ (the *triangle inequality*);

  ii. $p(\lambda x) = |\lambda| p(x)$ for all $x \in \mathcal{X}$ and $\lambda \in K$.

A *norm* on $\mathcal{X}$ is a seminorm that satisfies the additional property

  iii. $p(x) > 0$ for all $x \neq 0$.

Norms are generally denoted by $\|x\|$ rather than $p(x)$.

A *normed vector space* is a vector space $\mathcal{X}$ equipped with a norm $\| \cdot \|$. Every normed vector space is a metric space with the metric

$$\rho(x, y) = \|x - y\|.$$

A *Banach space* is a normed vector space that is complete with respect to this metric. Every normed vector space can be completed to make a Banach space; we shall exhibit an easy way to do this later in this section.

In a normed vector space one can consider convergence of sequences $\{x_n\}_1^\infty$ and series $\sum_1^\infty x_n$. A series $\sum_1^\infty x_n$ is called *absolutely convergent* if $\sum_1^\infty \|x_n\| < \infty$.

4.1 Proposition. *A normed vector space X is a Banach space if and only if every absolutely convergent series in X converges.*

Absolute convergence implies that the partial sums $S_N = \sum_1^N x_n$ are Cauchy, so completeness of X gives convergence of $\sum_1^\infty x_n$. On the other hand, if $\{y_n\}$ is a Cauchy sequence, one can find a subsequence $\{y_{n_j}\}$ so that $\|y_{n_j} - y_{n_{j-1}}\| \le 2^{-j}$. The telescoping series $y_{n_1} + \sum_2^\infty (y_{n_j} - y_{n_{j-1}})$ is then absolutely convergent, and its sum is $\lim_{n\to\infty} y_n$.

Here are some examples of Banach spaces:

- $\mathbb{R}^n$ or $\mathbb{C}^n$, with the Euclidean norm $\|x\| = (\sum_1^n |x_j|^2)^{1/2}$ (where $x = (x_1, \ldots, x_n)$). Sometimes it is more convenient to use the norm $\|x\| = \sum_1^n |x_j|$ or the norm $\|x\| = \max(|x_1|, \ldots, |x_n|)$; these all give equivalent metrics.

- The space $B(X)$ of all bounded real- or complex-valued functions on a set X, with the norm $\|f\| = \sup_{x \in X} |f(x)|$. If $\{f_n\}$ is Cauchy with respect to this norm, then $\{f_n(x)\}$ is Cauchy for each $x \in X$, so the completeness of $\mathbb{R}$ or $\mathbb{C}$ yields the existence of the limit.

- $L^1(\mu)$, where μ is any measure, with $\|f\| = \int |f| \, d\mu$. Here we must identify two functions that differ only on a set of measure zero. (We already considered the associated metric in Chapter 2.) Completeness follows from Proposition 4.1 and Theorem 2.13. (The space of Riemann integrable functions on $[a, b]$, with the norm $\|f\| = \int_a^b |f(x)| \, dx$, is not complete. This is one of the advantages of the Lebesgue integral.)

We shall consider other examples in Chapter 5.

A linear map $T : X \to Y$ between two normed vector spaces is said to be *bounded* if there is a constant $C \ge 0$ such that

$$\|Tx\| \le C\|x\| \text{ for all } x \in X.$$

4.2 Proposition. *If $T : X \to Y$ is a linear map between normed vector spaces, the following are equivalent:*
  a. T is continuous.
  b. T is continuous at 0.
  c. T is bounded.

The equivalence of (a) and (b) follows from the fact that T commutes with translations ($T(x + a) = Tx + Ta$). Moreover, (b) holds if and only if for every $\epsilon > 0$ there is a $\delta > 0$ such that $\|Tx\| < \epsilon$ whenever $\|x\| < \delta$. Since T commutes with scalar multiplication, this is equivalent to the existence of $c > 0$ such that $\|Tx\| \leq 1$ whenever $\|x\| \leq c$ and hence to the estimate $\|Tx\| \leq (1/c)\|x\|$, i.e., the boundedness of T.

If $\mathcal{X}$ and $\mathcal{Y}$ are normed vector spaces, we denote the space of all bounded linear maps from $\mathcal{X}$ to $\mathcal{Y}$ by $L(\mathcal{X}, \mathcal{Y})$. It is easy to check that $L(\mathcal{X}, \mathcal{Y})$ is a vector space, that the function $T \mapsto \|T\|$ defined by

$$\|T\| = \sup\{\|Tx\| : \|x\| = 1\} = \inf\{C : \|Tx\| \leq C\|x\| \text{ for all } x\}$$

is a norm on it (called the *operator norm*), and that $L(\mathcal{X}, \mathcal{Y})$ is a Banach space whenever $\mathcal{Y}$ is.

A particularly important case is where $\mathcal{Y}$ is the base field $\mathbb{F}$. A linear map from $\mathcal{X}$ to $\mathbb{F}$ is called a *linear functional* on $\mathcal{X}$; the space $L(\mathcal{X}, \mathbb{F})$ of all bounded linear functionals on $\mathcal{X}$ is called the *dual space* of $\mathcal{X}$ and is denoted by $\mathcal{X}^*$. By the preceding remarks, $\mathcal{X}^*$ is a Banach space with the operator norm, whether $\mathcal{X}$ is a Banach space or not.

Here is the fundamental existence theorem for bounded linear functionals on normed vector spaces, and more generally for linear functionals bounded by seminorms.

4.3 The Hahn-Banach theorem. *Let $\mathcal{X}$ be a vector space over $\mathbb{F}$, p a seminorm on $\mathcal{X}$, $\mathcal{V}$ a vector subspace of $\mathcal{X}$, and f a linear functional on $\mathcal{V}$ such that $|f(x)| \leq p(x)$ for $x \in \mathcal{V}$. Then there is a linear functional F on $\mathcal{X}$ such that $|F(x)| \leq p(x)$ for all $x \in \mathcal{X}$ and $F|\mathcal{V} = f$.*

To prove this for $\mathbb{F} = \mathbb{R}$, one first verifies, by some elementary calculations, that it is always possible to extend f from $\mathcal{V}$ to a space containing one additional vector $x \notin \mathcal{V}$ (i.e., to the linear span of $\mathcal{V}$ and x) while maintaining the estimate $|f| \leq p$. One then considers the family of all extensions g of f to linear functionals on larger subspaces of $\mathcal{X}$ satisfying $|g| \leq p$; by Zorn's lemma, this family has a maximal element, and the preceding argument shows that the domain of this maximal element must be all of $\mathcal{X}$. The case $\mathbb{F} = \mathbb{C}$ can be reduced to the case $\mathbb{F} = \mathbb{R}$ by using the fact that a complex-linear functional f can be recovered from its real part $u = \text{Re } f$ by the formula $f(x) = u(x) - iu(ix)$.

If $\mathcal{X}$ is a normed vector space, the Hahn-Banach theorem with $p(x) = \|x\|$ immediately yields the following useful results:

- If $x \in \mathcal{X} \setminus \{0\}$, there exists $f \in \mathcal{X}^*$ such that $\|f\| = 1$ and $f(x) = \|x\|$.

- If W is a closed subspace of X and $x \notin W$, there exists $f \in X^*$ such that $f = 0$ on W and $f(x) = 1$. (Apply the Hahn-Banach theorem with $V = $ the linear span of W and x, $f(w + \lambda x) = \lambda$ for $w \in W$, and $p(y) = C^{-1}\|y\|$ where C is the distance from x to W.)

If X is a normed vector space, every $x \in X$ defines a bounded linear functional $\widehat{x}$ on X^* by $\widehat{x}(f) = f(x)$; this is the "duality" between X and X^*. Since $|\widehat{x}(f)| \leq \|f\| \|x\|$, with equality if f is chosen to satisfy $\|f\| = 1$ and $f(x) = \|x\|$, we see that the operator norm $\|\widehat{x}\|$ is equal to $\|x\|$. Thus, the map $x \mapsto \widehat{x}$ is an isometric embedding of X into X^{**}.

Since X^{**} is always a Banach space, the closure of $\widehat{X} = \{\widehat{x} : x \in X\}$ in X^{**} is a Banach space into which X is embedded as a dense subspace; it is called the *completion* of X. Of course, if X is itself a Banach space, then $\widehat{X}$ is already closed in X^{**}. When X is finite-dimensional, we always have $\widehat{X} = X^{**}$ since these spaces have the same dimension, but in infinite dimensions $\widehat{X}$ is usually a proper subspace of X^{**}. A Banach space X such that $\widehat{X} = X^{**}$ is said to be *reflexive*. We shall produce some examples of reflexive and nonreflexive spaces later on.

We conclude this section with a group of fundamental theorems about bounded linear maps between Banach spaces.

4.4 The open mapping theorem. *Let X and Y be Banach spaces. If $T \in L(X, Y)$ is surjective, then $T(U)$ is open in Y whenever U is open in X.*

Let B_r denote the open ball of radius r about 0 in X. Since T commutes with translations and scalar multiplication, the proof reduces to showing that $T(B_1)$ contains a ball about 0 in Y. Since T is surjective, we have $Y = \bigcup_1^\infty T(B_n)$. Since Y is complete and $T(B_n) = \{ny : y \in T(B_1)\}$, Corollary 1.17 implies that the closure $\overline{T(B_1)}$ has nonempty interior. Since this set is convex and symmetric about the origin in Y, it follows without difficulty that it contains a ball $B(r, 0) \subset Y$. Finally, one uses the completeness of X to show that $T(B_1)$ contains the ball $B(r/2, 0) \subset Y$.

The most important case of the open mapping theorem is where T is bijective. In this case, the fact that T maps open sets to open sets means that the inverse map T^{-1} is continuous. Hence:

4.5 Corollary. *If X and Y are Banach spaces and $T \in L(X, Y)$ is bijective, then $T^{-1} \in L(Y, X)$.*

If X and Y are normed vector spaces, a linear map $T : X \to Y$ is called *closed* if its graph

$$\Gamma(T) = \{(x, y) \in X \times Y : y = Tx\}$$

is a closed subspace of $\mathcal{X} \times \mathcal{Y}$ (which is a normed linear space with norm $\|(x, y)\| = \|x\| + \|y\|$).

4.6 The closed graph theorem. *Let $\mathcal{X}$ and $\mathcal{Y}$ be Banach spaces. If $T : \mathcal{X} \to \mathcal{Y}$ is a closed linear map, then T is bounded.*

Indeed, since $\mathcal{X}$ and $\mathcal{Y}$ are complete, so is $\mathcal{X} \times \mathcal{Y}$, and hence so is $\Gamma(T)$ when T is closed. The projections $\pi_1(x, Tx) = x$ and $\pi_2(x, Tx) = Tx$ from $\Gamma(T)$ to $\mathcal{X}$ and $\mathcal{Y}$ are obviously bounded linear maps, and π_1 is a bijection. By Corollary 4.5, π_1^{-1} is bounded, and hence so is $T = \pi_2 \circ \pi_1^{-1}$.

Let us pause to examine the meaning of the closed graph theorem. The closedness of T means that if $(x_n, Tx_n) \to (x, y)$ then $y = Tx$; that is, if $x_n \to x$ and $Tx_n \to y$ then $y = Tx$. The boundedness or continuity of T, on the other hand, means that if $x_n \to x$ then $Tx_n \to Tx$. Thus the point of the closed graph theorem is that in verifying this last condition one is allowed to assume that $\{Tx_n\}$ is convergent, and one needs only to show that the limit is the right thing, namely, Tx.

It should be emphasized that the completeness of $\mathcal{X}$ and $\mathcal{Y}$ is essential for the open mapping and closed graph theorems. Closed but unbounded linear maps $T : \mathcal{X} \to \mathcal{Y}$ are commonplace when $\mathcal{X}$ is not complete. Such maps play an important role in the applications of functional analysis to differential equations, since differential operators tend to be unbounded, and to quantum physics, where they represent observable quantities whose range of possible values is unbounded.

Our final theorem permits one to pass from pointwise estimates to uniform estimates in certain situations.

4.7 The uniform boundedness principle. *Suppose $\mathcal{X}$ and $\mathcal{Y}$ are Banach spaces and $\mathcal{A} \subset L(\mathcal{X}, \mathcal{Y})$. If $\sup_{T \in \mathcal{A}} \|Tx\| < \infty$ for all $x \in \mathcal{X}$, then $\sup_{T \in \mathcal{A}} \|T\| < \infty$.*

The quickest proof proceeds by applying Corollary 1.17 to the decomposition $\mathcal{X} = \bigcup_1^\infty F_n$ where $F_n = \{x : \sup_{T \in \mathcal{A}} \|Tx\| \leq n\}$. There is also a neat and more elementary proof that does not use the Baire category theorem; see Hennefeld [8]. Here is a typical application of the uniform boundedness principle:

4.8 Corollary. *Suppose $\mathcal{X}$ is a Banach space and E is a subset of $\mathcal{X}$. If $f(E)$ is a bounded subset of the base field $\mathbb{F}$ for every $f \in \mathcal{X}^*$, then E itself is bounded in $\mathcal{X}$.*

To prove this, identify $\mathcal{X}$ with $\widehat{\mathcal{X}} \subset \mathcal{X}^{**}$ and apply Theorem 4.7 with $\mathcal{X}^*$ and $\mathbb{F}$ playing the roles of $\mathcal{X}$ and $\mathcal{Y}$.

4.2 HILBERT SPACES

Let $\mathcal{H}$ be a complex vector space. An *inner product* on $\mathcal{H}$ is a map $(x, y) \mapsto \langle x, y \rangle$ from $\mathcal{H} \times \mathcal{H}$ to $\mathbb{C}$ that is

   i. *linear in the first variable* $\left(\langle c_1 x_1 + c_2 x_2, y \rangle = c_1 \langle x_1, y \rangle + c_2 \langle x_2, y \rangle \right)$,

   ii. *Hermitian* $\left(\langle y, x \rangle = \overline{\langle x, y \rangle} \right)$, and

   iii. *positive definite* $\left(\langle x, x \rangle > 0 \text{ for all } x \neq 0 \right)$.

The first two properties imply that inner products are conjugate-linear in the second variable, that is, $\langle x, c_1 y_1 + c_2 y_2 \rangle = \overline{c}_1 \langle x, y_1 \rangle + \overline{c}_2 \langle x, y_2 \rangle$. This combination of linearity and conjugate-linearity in the two variables is sometimes called *sesquilinearity*. (In the physics literature, inner products are taken to be linear in the second variable and conjugate-linear in the first, and they are commonly denoted by $\langle x | y \rangle$.)

A basic property of inner products is the estimate

$$(4.9) \qquad\qquad |\langle x, y \rangle|^2 \leq \langle x, x \rangle \langle y, y \rangle,$$

variously known as *Cauchy's inequality*, *Buniakovsky's inequality*, or the *Schwarz inequality*. It is trivial if $y = 0$; otherwise, let $z = e^{i\theta} y$ where θ is chosen so that $\langle x, z \rangle = e^{-i\theta} \langle x, y \rangle$ is real and positive. We have

$$0 \leq \langle x - tz, x - tz \rangle = \langle x, x \rangle - 2t \langle x, z \rangle + t^2 \langle z, z \rangle$$

for all real t, and in particular for the t that minimizes the quadratic function on the right, namely, $t = \langle x, z \rangle / \langle z, z \rangle = |\langle x, y \rangle| / \langle y, y \rangle$. Substituting this value for t yields the desired result.

The inequality (4.9) shows that for any x and y we have

$$\langle x + y, x + y \rangle = \langle x, x \rangle + 2 \operatorname{Re}\langle x, y \rangle + \langle y, y \rangle \leq \left(\langle x, x \rangle + \langle y, y \rangle \right)^2,$$

and it follows that the function

$$\|x\| = \sqrt{\langle x, x \rangle}$$

is a norm on $\mathcal{H}$. A complex vector space equipped with an inner product that is complete with respect to the associated norm is called a *Hilbert space*.

A simple example of a Hilbert space (the one originally considered by Hilbert, in fact) is the space l^2 of square-summable sequences of complex numbers:

$$l^2 = \left\{ \{c_j\}_1^\infty : \sum_1^\infty |c_j|^2 < \infty \right\},$$

with inner product $\langle \{c_j\}, \{d_j\} \rangle = \sum_1^\infty c_j \overline{d}_j$. (The absolute convergence of this series follows from the fact that $2ab \le a^2 + b^2$ for all $a, b \ge 0$.) The proof of completeness is an easy exercise. We shall produce other examples of Hilbert spaces in §5.1.

Henceforth we assume that $\mathcal{H}$ is a Hilbert space. If $x, y \in \mathcal{H}$, we say that x is *orthogonal* to y and write $x \perp y$ if $\langle x, y \rangle = 0$. If $x_1, \ldots, x_n$ are pairwise orthogonal, we have the "Pythagorean theorem"

$$(4.10) \qquad \left\| \sum_1^n x_j \right\|^2 = \sum_1^n \|x_j\|^2,$$

because $\| \sum x_j \|^2 = \sum_{j,k} \langle x_j, x_k \rangle$ and all the terms with $j \ne k$ vanish.

For $E \subset \mathcal{H}$, the *orthogonal complement* of E is

$$E^\perp = \{x \in \mathcal{H} : \langle x, y \rangle = 0 \text{ for all } y \in E\}.$$

It is easily verified that $E^\perp$ is always a closed subspace of $\mathcal{H}$. The basic fact about orthogonal complements is the following.

4.11 Proposition. *If $\mathcal{V}$ is a closed subspace of $\mathcal{H}$, then $\mathcal{H} = \mathcal{V} \oplus \mathcal{V}^\perp$; that is, every $x \in \mathcal{H}$ is uniquely expressible as $x = y + z$ where $y \in \mathcal{V}$ and $z \in \mathcal{V}^\perp$.*

Uniqueness is clear since $\mathcal{V} \cap \mathcal{V}^\perp = \{0\}$. The geometric intuition is that y and z are the orthogonal projections of x onto $\mathcal{V}$ and $\mathcal{V}^\perp$, which are the elements of $\mathcal{V}$ and $\mathcal{V}^\perp$ that are closest to x. In fact, one shows that if $\delta = \inf\{\|x - y\| : y \in \mathcal{V}\}$ and $\{y_n\}$ is any sequence in $\mathcal{V}$ such that $\|x - y_n\| \to \delta$, then $\{y_n\}$ is Cauchy. It follows that there is a unique $y \in \mathcal{V}$ that is closest to x, and one can then verify that $x - y \in \mathcal{V}^\perp$.

This proposition leads to a neat description of the dual space $\mathcal{H}^*$:

4.12 Theorem. *For every $f \in \mathcal{H}^*$ there is a unique $y \in \mathcal{H}$ such that $f(x) = \langle x, y \rangle$; moreover, $\|y\| = \|f\|$.*

This is obvious if f is the zero functional. Otherwise, let $\mathcal{V}$ be the nullspace of f; since $\mathcal{V} \ne \mathcal{H}$, there is a nonzero $z \in \mathcal{V}^\perp$. For any $x \in \mathcal{H}$ the vector $u = f(x)z - f(z)x$ belongs to $\mathcal{V}$, so $0 = \langle u, z \rangle = f(x)\|z\|^2 - f(z)\langle x, z \rangle$; thus $f(x) = \langle x, y \rangle$ where $y = \overline{f(z)}z/\|z\|^2$. The fact that the operator norm $\|f\|$ equals the $\mathcal{H}$-norm $\|y\|$ comes from inequality (4.9), which is an equality when $x = y$. For uniqueness, if also $f(x) = \langle x, y' \rangle$, by taking $x = y - y'$ one sees that $\|y - y'\|^2 = 0$ and hence $y = y'$.

The correspondence $f \leftrightarrow y$ in Theorem 4.12 is a conjugate-linear isometric bijection between $\mathcal{H}$ and $\mathcal{H}^*$. Hence $\mathcal{H}^*$ is also a Hilbert space, and $\mathcal{H}$ is reflexive.

A sequence $\{u_j\}_1^\infty$ in $\mathcal{H}$ is said to be *orthonormal* if $\|u_j\| = 1$ for all α and $\langle u_j, u_k \rangle = 0$ whenever $j \neq k$. If $\{u_j\}$ is orthonormal and $\{c_j\}$ is a sequence of complex numbers, by (4.10) we have $\| \sum_m^n c_j u_j \|^2 = \sum_m^n |c_j|^2$, so the partial sums of the series $\sum c_j u_j$ are Cauchy if and only if those of $\sum |c_j|^2$ are. Since $\mathcal{H}$ is complete, we see that $\sum c_j u_j$ converges (in the norm topology of $\mathcal{H}$) if and only if $\sum |c_j|^2 < \infty$, in which case (4.10) remains true in the limit: $\| \sum_1^\infty c_j u_j \|^2 = \sum_1^\infty |c_j|^2$.

4.13 Theorem. *If $\{u_n\}_1^\infty$ is an orthonormal sequence in $\mathcal{H}$, the following are equivalent:*

   a. *(Completeness) If $\langle x, u_j \rangle = 0$ for all j, then $x = 0$.*
   b. *(Parseval's identity) $\|x\|^2 = \sum_1^\infty |\langle x, u_j \rangle|^2$ for all $x \in \mathcal{H}$.*
   c. *$x = \sum_1^\infty \langle x, u_j \rangle u_j$ for all $x \in \mathcal{H}$.*

The key to the proof is the fact that for any $x \in \mathcal{H}$,

$$\left\| x - \sum_1^n \langle x, u_j \rangle u_j \right\|^2 = \|x\|^2 - \sum_1^n |\langle x, u_j \rangle|^2,$$

which follows from a calculation involving (4.10). Since the quantity on the left is nonnegative, we have $\sum_1^n |\langle x, u_j \rangle|^2 \leq \|x\|^2$ for all n, so the series $\sum_1^\infty |\langle x, u_j \rangle|^2$ converges. Hence, by the remarks preceding the theorem, the series $\sum \langle x, u_j \rangle u_j$ converges, and if (c) holds then so does (b). Moreover, (b) obviously implies (a). Finally, let $y = x - \sum_1^\infty \langle x, u_j \rangle u_j$. We have $\langle y, u_k \rangle = \langle x, u_k \rangle - \langle x, u_k \rangle = 0$ for all k, so if (a) holds then $y = 0$, whence (c) holds.

An orthonormal sequence $\{u_j\}$ that possesses the properties (a)–(c) in Theorem 4.13 is called an *orthonormal basis*. A Hilbert space $\mathcal{H}$ possesses an orthonormal basis in this sense if and only if it is separable, as the finite sums $\sum_1^n c_j u_j$ with $\operatorname{Re} c_j$ and $\operatorname{Im} c_j$ rational form a countable dense set in $\mathcal{H}$. (Conversely, if $\{x_j\}$ is a countable dense set, one can turn it into a countable orthonormal basis by a standard algorithm of linear algebra, the Gram-Schmidt process.) The preceding results generalize to inseparable spaces by considering orthonormal sets $\{u_\alpha\}_{\alpha \in A}$ with uncountable index sets A, although the apparently uncountable sums such as $\sum_{\alpha \in A} |\langle x, u_\alpha \rangle|^2$ require some explanation. However, almost all the Hilbert spaces that one encounters in practice are separable.

We conclude with describing some particularly important types of linear maps on Hilbert spaces. Suppose $\mathcal{H}_1$ and $\mathcal{H}_2$ are Hilbert spaces; we denote the inner products and norms on both spaces by $\langle \cdot, \cdot \rangle$ and $\| \cdot \|$. If $T \in L(\mathcal{H}_1, \mathcal{H}_2)$ and $y \in \mathcal{H}_2$, the map $x \to \langle Tx, y \rangle$ is a linear functional on $\mathcal{H}_1$, and by (4.9) it is bounded:

$$|\langle Tx, y \rangle| \leq \|Tx\| \, \|y\| \leq \|T\| \, \|y\| \, \|x\|.$$

Hence by Theorem 4.12 there is a unique element of $\mathcal{H}_1$, which we denote by T^*y, such that $\langle Tx, y \rangle = \langle x, T^*y \rangle$ for all x; moreover, $\|T^*y\| \leq \|T\| \, \|y\|$. The map $T^* : \mathcal{H}_2 \to \mathcal{H}_1$ thus defined is a bounded linear map from $\mathcal{H}_2$ to $\mathcal{H}_1$; it is called the *adjoint* of T. The preceding inequality shows that $\|T^*\| \leq \|T\|$, and since $(T^*)^* = T$, we also have $\|T\| \leq \|T^*\|$, so $\|T^*\| = \|T\|$. When $\mathcal{H}_1 = \mathcal{H}_2$, a bounded linear map T such that $T = T^*$ is called *self-adjoint* or *Hermitian*.

A linear map $U : \mathcal{H}_1 \to \mathcal{H}_2$ is called *unitary* if it is bijective and preserves inner products, that is, $\langle Ux_1, Ux_2 \rangle = \langle x_1, x_2 \rangle$ for all $x_1, x_2 \in \mathcal{H}_1$. (The latter condition always implies that U is injective, but in infinite dimensions surjectivity must be assumed separately.) Equivalently, an invertible map $U \in L(\mathcal{H}_1, \mathcal{H}_2)$ is unitary if and only if $U^{-1} = U^*$. For example, by Theorem 4.13 and the remarks preceding it, for any orthonormal basis $\{u_j\}_1^\infty$ for $\mathcal{H}$, the map that takes $x \in \mathcal{H}$ to the sequence $\{\langle x, u_j \rangle\}$ is unitary from $\mathcal{H}$ to l^2.

4.3 OTHER TOPOLOGICAL VECTOR SPACES

On some vector spaces the natural topological structure is given not by a single norm but by families of norms or seminorms. To be precise, let $\mathcal{X}$ be a vector space, and let $\{p_\alpha\}_{\alpha \in A}$ be a family of seminorms on $\mathcal{X}$. The topology on $\mathcal{X}$ generated by the "balls" $\{x \in \mathcal{X} : p_\alpha(x - x_0) < r\}$ ($x_0 \in \mathcal{X}$, $\alpha \in A$, and $r > 0$) is called the topology *generated by*, or *associated to*, the family $\{p_\alpha\}_{\alpha \in A}$. Usually one wants such topologies to be Hausdorff; this happens precisely when for every $x \neq 0$ there is an α such that $p_\alpha(x) \neq 0$. Otherwise, the set $\mathcal{N} = \{x : p_\alpha(x) = 0 \text{ for all } \alpha\}$ is a vector subspace of $\mathcal{X}$, and one can consider the quotient space $\mathcal{X}/\mathcal{N}$ instead. The seminorms p_α induce seminorms on this space, and the associated topology on it is Hausdorff.

Here are some basic facts about topologies defined by families of seminorms; their verification is routine.

4.14 Proposition. *Let X be a vector space equipped with a family $\{p_\alpha\}_{\alpha \in A}$ of seminorms and its associated topology.*

a. *The vector operations are continuous; that is, the map $(x, y) \mapsto x + y$ is continuous from $X \times X$ to X, and the map $(c, x) \mapsto cx$ is continuous from $\mathbb{F} \times X$ to X.*

b. *If $\langle x_i \rangle_{i \in I}$ is a net in X, then $x_i \to x$ if and only if $p_\alpha(x_i - x) \to 0$ for all $\alpha \in A$.*

c. *Suppose Y is another vector space equipped with a family of seminorms $\{q_\beta\}_{\beta \in B}$. A linear map $T : X \to Y$ is continuous if and only if for every $\beta \in B$ there exist $C > 0$ and a finite $F \subset A$ such that $q_\beta(Tx) \leq C \sum_{\alpha \in F} p_\alpha(x)$ for all $x \in X$.*

There is an extensive theory of vector spaces with topologies defined by families of seminorms, but it is outside the scope of this book. Here we only give a brief description of some of the examples that are most commonly encountered in practice. First, here are two specific examples.

- Let X be an LCH space, and let $X = C(X)$ be the space of continuous complex-valued functions on X. The *topology of uniform convergence on compact sets* is the topology defined by the seminorms $p_K(f) = \sup_{x \in K} |f(x)|$ as K ranges over all compact subsets of X. This topology is of great importance in many parts of analysis, especially complex variable theory; we shall say more about it in §5.2.

- Let $X = C^\infty([0, 1])$ be the space of functions on $[0, 1]$ that possess derivatives of all orders on $[0, 1]$ (including one-sided derivatives at the endpoints), and define the seminorms p_k on X for $k = 0, 1, 2, \ldots$ by $p_k(f) = \sup_{t \in [0,1]} |f^{(k)}(t)|$. The associated topology on X is called the C^∞ *topology*. The linear operator d/dt on X is continuous with respect to this topology, almost by definition. In contrast, there is no norm on X with respect to which d/dt is bounded, for every complex number λ is an eigenvalue (with eigenfunction $e^{\lambda t}$).

The other common examples of topologies defined by families of seminorms arise from the following situation. Suppose X is a vector space, Y is a normed vector space, and $\{T_\alpha\}_{\alpha \in A}$ is family of linear maps from X to Y. For each α, the function $p_\alpha(x) = \|T_\alpha x\|$ is a seminorm on x (it is a norm precisely when T is injective). The topology defined by these seminorms is easily seen to be the weakest topology on X such that the maps T_α are all continuous; we call it the topology *generated by* the family $\{T_\alpha\}_{\alpha \in A}$. Here are the two most basic classes of examples.

- Let X be a normed vector space. The topology on X generated by X^* (a family of maps from X to the base field $\mathbb{R}$ or $\mathbb{C}$) is called the *weak topology* on X. The weak topology is weaker than the norm topology, strictly so unless X is finite-dimensional. A net $\langle x_\alpha \rangle$ in X converges to x in the weak topology if and only if $f(x_\alpha) \to f(x)$ for all $f \in X^*$.

- Let X be a normed vector space and X^* its dual space. The *weak* topology* (read "weak star topology") on X^* is the topology generated by the evaluation maps $f \mapsto f(x)$ for $x \in X$, that is, the topology generated by X considered as a subspace of X^{**}. Convergence in the weak* topology is pointwise convergence: $f_\alpha \to f$ (weak*) means that $f_\alpha(x) \to f(x)$ for all x. The weak* topology is weaker than the weak topology on X^* as defined in the preceding item, strictly so unless X is reflexive.

An example: An orthonormal basis $\{e_j\}_1^\infty$ for a Hilbert space $\mathcal{H}$ does not converge in the norm topology because $\|e_j - e_k\| = \sqrt{2}$ for any $j \neq k$, but $e_j \to 0$ weakly. This follows from Theorems 4.12 and 4.13: the convergence of $\sum |\langle e_j, x \rangle|^2$ implies that $\langle e_j, x \rangle \to 0$, for any $x \in \mathcal{H}$.

The weak* topology on a dual space X^* is particularly important because of the following result.

4.15 Alaoglu's theorem. *For any normed vector space X, the closed unit ball $B^* = \{f \in X^* : \|f\| \leq 1\}$ in X^* is compact in the weak* topology.*

The proof is quite neat. For each $x \in X$, let $D_x = \{t \in \mathbb{F} : |t| \leq \|x\|\}$, and let $D = \prod_{x \in X} D_x$. D is the set of $\mathbb{F}$-valued functions f on X that satisfy $|f(x)| \leq \|x\|$ for all x, and B^* is the subset consisting of linear functions. The topologies on B^* inherited from the weak* topology on X^* and the product topology on D are the same, namely the topology of pointwise convergence. Moreover, B^* is closed in D because the pointwise limit of linear functions is linear. But D is compact by Tychonoff's theorem, so the result follows immediately.

CHAPTER 5

FUNCTION SPACES

In this chapter we study some of the spaces of functions that are of fundamental importance in modern analysis.

5.1 L^p SPACES

Let $(X, \mathcal{M}, \mu)$ be a measure space. We recall that $L^1(\mu)$ is the space of all μ-integrable complex-valued functions on X. For $0 < p < \infty$, we define the space $L^p(\mu)$ (also denoted by $L^p(X)$ or simply L^p when μ is understood) to be the set of all measurable complex-valued functions f on X such that $|f|^p \in L^1(\mu)$. Thus, with the notation

$$(5.1) \qquad \|f\|_p = \left[\int |f|^p \, d\mu \right]^{1/p},$$

$L^p(\mu)$ is the space of all measurable functions on X such that $\|f\|_p < \infty$. As in the case of $L^1(\mu)$, two functions are considered to define the same element of $L^p(\mu)$ if they are equal almost everywhere.

We proceed to develop the theory of L^p spaces on a fixed measure space $(X, \mathcal{M}, \mu)$, which is based on the following close relative of the Cauchy-Schwarz inequality. In it, and throughout this section, two numbers $p, q \in (1, \infty)$ are said to be *conjugate* to each other if

$$(5.2) \qquad \frac{1}{p} + \frac{1}{q} = 1.$$

5.3 Hölder's inequality. *Suppose $p, q \in (1, \infty)$ are conjugate to each other. If f and g are measurable functions on X, then*

$$\|fg\|_1 \le \|f\|_p \|g\|_q.$$

In particular, if $f \in L^p$ and $g \in L^q$, then $fg \in L^1$.

By multiplying f and g by scalars, one reduces this to the case where $\|f\|_p = \|g\|_q = 1$. In that case the result follows by applying the elementary inequality

$$a^\lambda b^{1-\lambda} \le \lambda a + (1 - \lambda)b \qquad (a, b \ge 0; \; 0 < \lambda < 1)$$

to $a = |f(x)|^p, b = |g(x)|^q$, and $\lambda = 1/p$, and integrating over X.

We can extend the relation (5.2) to the extreme case $q = 1$ by declaring 1 and ∞ to be conjugate to each other, and it is important to extend the notion of L^p space to the limiting case $p = \infty$ in such a way that Hölder's inequality remains valid. That is, we want L^∞ to be the space of measurable functions f on X such that $fg \in L^1$ for all $g \in L^1$, and the norm $\|f\|_\infty$ to be defined so that $\|fg\|_1 \le \|f\|_\infty \|g\|_1$. A little thought should convince one that L^∞ should be the space of all bounded measurable functions on X, and the norm $\|f\|_\infty$ should be the supremum of $|f|$ over X — except that one wants to be able to neglect the behavior of functions on sets of measure zero. The precise definition is as follows:

$$L^\infty(\mu) = \{\text{measurable } f : X \to \mathbb{C} : \|f\|_\infty < \infty\},$$

where

$$\|f\|_\infty = \inf\{a \ge 0 : \mu(\{x : |f(x)| > a\}) = 0\}.$$

The quantity $\|f\|_\infty$ is sometimes called the *essential supremum* of $|f|$ and written as

$$\|f\|_\infty = \text{ess sup}_{x \in X} |f(x)|.$$

(Another justification for this definition is that if $f \in L^p$ for all $p > p_0$, then $\|f\|_\infty = \lim_{p \to \infty} \|f\|_p$, the proof of which is an instructive exercise.)

The notation $\|f\|_p$ suggests that $\| \cdot \|_p$ is a norm. It is obvious that $\|cf\|_p = |c| \|f\|_p$ for all scalars c and that $\|f\|_p = 0$ only when $f = 0$ (almost everywhere), so the main question is the triangle inequality.

5.4 Minkowski's inequality. *Suppose $1 \le p \le \infty$. If $f, g \in L^p$, then*

$$\|f + g\|_p \le \|f\|_p + \|g\|_p.$$

This is obvious when $p = 1$ or $p = \infty$. For $1 < p < \infty$, it is proved by noting that $|f + g|^p \le (|f| + |g|)|f + g|^{p-1}$ and applying Hölder's inequality to the products $|f||f + g|^{p-1}$ and $|g||f + g|^{p-1}$.

In short, when $1 \leq p \leq \infty$, L^p is a normed vector space with norm $\| \cdot \|_p$, and we shall restrict our attention to this case from now on. (When $p < 1$, it is not $\| \cdot \|_p$ but $\| \cdot \|_p^p$ that satisfies the triangle inequality and makes L^p into a metric space. L^p spaces with $p < 1$ are useful in certain contexts, but we shall say no more about them here.) Convergence with respect to the L^1 norm is the "convergence in L^1" discussed in §2.3, and convergence with respect to the L^p norm for $1 < p < \infty$ is rather similar. It is an easy exercise to see that convergence in the L^∞ norm is uniform convergence except on a set of measure zero (that is, $\| f_n - f \|_\infty \to 0$ if and only if there is a set $E \subset X$ with $f_n \to f$ uniformly on E and $\mu(X \setminus E) = 0$).

5.5 Proposition. *For $1 \leq p \leq \infty$, L^p is a Banach space.*

For $p = \infty$ this is an easy consequence of the preceding remark. For $1 \leq p < \infty$, one uses Proposition 4.1. The case $p = 1$ follows immediately from Theorem 2.13, and the case $1 < p < \infty$ is similar. That is, if $\{f_n\} \subset L^p$ and $\sum \| f_n \|_p < \infty$, one shows that $\sum |f_n| \in L^p$, and in particular $\sum |f_n| < \infty$ almost everywhere, and hence that $\sum f_n$ converges almost everywhere and in the L^p norm.

The case $p = 2$ is special. It follows from Hölder's inequality that if $f, g \in L^2$ then $f \overline{g} \in L^1$, so the formula

$$\langle f, g \rangle = \int f \overline{g} \, d\mu$$

defines an inner product on L^2, and the associated norm is the L^2 norm. In view of Proposition 5.5, we see that L^2 is a Hilbert space.

A few remarks are in order concerning the meaning of the condition $f \in L^p$ and the relationships between different L^p spaces. To begin with, we observe that for any $a > 0$,

$$\int_X |f|^p \, d\mu \geq \int_{\{x : |f(x)| > a\}} |f|^p \, d\mu \geq a^p \mu(\{x : |f(x)| > a\}),$$

or in other words,

$$(5.6) \qquad \mu(\{x : |f(x)| > a\}) \leq \left[\frac{\|f\|_p}{a} \right]^p.$$

This is known as *Chebyshev's inequality*; it places a restriction on the size of the sets on which an L^p function can be larger than some specified amount.

Roughly speaking, a function f can fail to be in L^p either because it blows up too rapidly near some point(s), so that $\mu(\{x : |f(x)| > a\})$

becomes too big as $a \to \infty$, or because it decays too slowly at infinity, so that $\mu(\{x : |f(x)| > a\})$ becomes too big as $a \to 0$. Raising to a high power makes large numbers larger and small numbers smaller, so when $p < q$, functions in L^p can have worse singularities than functions in L^q, but functions in L^q can decay more slowly at infinity. A typical class of examples is provided by $f_a(x) = x^{-a}$ on $(0, \infty)$ (with Lebesgue measure): we have $f_a \in L^p((0, 1])$ if and only if $ap < 1$, but $f_a \in L^p([1, \infty))$ if and only if $ap > 1$.

If $\mu(X) < \infty$ so that there is no question of "decay at infinity," we have $L^p \supset L^q$ for $p < q$. On the other hand, if there are no subsets of X of arbitrarily small positive measure (for example, if μ is counting measure) so that L^p functions must be bounded (by Chebyshev's inequality), we have $L^p \subset L^q$ for $p < q$. In general, for $1 \le p < q < r \le \infty$ we always have

$$L^p \cap L^r \subset L^q \subset L^p + L^r,$$

where $L^p + L^r = \{f + g : f \in L^p, \, g \in L^r\}$. The proofs of these assertions are all easy.

Here are some useful approximation results:

5.7 Proposition. *Let $(X, \mathcal{M}, \mu)$ be a measure space.*

a. *The simple functions are dense in $L^\infty(\mu)$.*

b. *The simple functions that vanish outside sets of finite measure are dense in $L^p(\mu)$ for $p < \infty$.*

c. *If X is an LCH space and μ is a regular Borel measure, then $C_c(X)$ is dense in $L^p(\mu)$ for $p < \infty$.*

The first two assertions follow easily from Proposition 2.6, and the last one is proved in the same way as Proposition 3.27 (the case $p = 1$).

We now turn to the duality theory of L^p spaces. If p and q are conjugate exponents, Hölder's inequality shows that every $g \in L^q$ defines a bounded linear functional ϕ_g on L^p by

$$(5.8) \qquad\qquad \phi_g(f) = \int fg \, d\mu$$

and that the norm of ϕ_g in $(L^p)^*$ is at most $\|g\|_q$. In fact, the norm of ϕ_g is exactly $\|g\|_q$ when $q < \infty$, as one sees by taking $f = |g|^q/g$ with the understanding that $f(x) = 0$ wherever $g(x) = 0$. The same result holds for $q = \infty$ provided μ is σ-finite, as one sees by taking $f = \chi_E |g|/g$ where E is a set of finite positive measure on which $|g| > \|g\|_\infty - \epsilon$. The main result is that in most cases, every element of $(L^p)^*$ is of this form.

5.9 Theorem. *Suppose μ is σ-finite, $1 \le p < \infty$, and p and q are conjugate to each other. Then the map $g \mapsto \phi_g$ defined by (5.8) is a norm-preserving bijection from L^q to $(L^p)^*$. In particular, L^p is reflexive for $1 < p < \infty$.*

To prove this we need to show that every $\phi \in (L^p)^*$ is of the form ϕ_g, and the idea is as follows. First suppose that μ is a finite measure, so that the characteristic function of every measurable set is in L^p, and define $\nu(E) = \phi(\chi_E)$. If $E = \bigcup_1^\infty E_n$ where the E_n's are disjoint, we have $\chi_E = \sum_1^\infty \chi_{E_n}$ where the series converges in the L^p norm (this is where we need the assumption that $p < \infty$), and it follows that ν is a complex measure on $(X, \mathcal{M})$ that is absolutely continuous with respect to μ. Hence, by the Radon-Nikodym theorem we have $d\nu = g \, d\mu$ for some $g \in L^1(\mu)$, and $\phi(f) = \int fg \, d\mu$ for all simple functions f. One uses the estimate $|\int fg \, d\mu| \le \|\phi\| \, \|f\|_p$ to show that $g \in L^q$; it then follows that $\phi(f) = \int fg \, d\mu$ for all $f \in L^p$. If μ is merely σ-finite, one writes X as a countable union of sets of finite measure, applies this argument on each of these sets, and patches the results together.

An additional argument can be adduced to show that Theorem 5.9 is valid for $1 < p < \infty$ with no restriction on μ.

As for the case $p = \infty$, the remarks preceding Theorem 5.9 show that the map $g \mapsto \phi_g$ is always a norm-preserving injection of L^1 into $(L^\infty)^*$, but it is usually not a surjection. There are various ways of producing examples of linear functionals on L^∞ that do not come from an element of L^1; here is one. On $\mathbb{R}$ with Lebesgue measure λ, let ϕ_n be the element of $(L^\infty)^*$ defined by the function $\chi_{[-n,n]}/2n \in L^1$:

$$\phi_n(f) = \frac{1}{2n} \int_{[-n,n]} f \, d\lambda.$$

Then $\|\phi_n\| = \|\chi_{[-n,n]}\|_1/2n = 1$ for all n. By Alaoglu's theorem, the sequence $\{\phi_n\}$ has a weak* cluster point. It is not zero because it takes the value 1 on the constant function 1 (as all the ϕ_n do), but it annihilates all functions that vanish outside a bounded set; hence it cannot be given by integration against an L^1 function.

We conclude this section with a neat application of the duality theorem. Minkowski's inequality says that the L^p norm of a sum is at most the sum of the L^p norms; we now show that an analogous result holds with sums replaced by integrals.

5.10 Minkowski's inequality for integrals. *Let $(X, \mathcal{M}, \mu)$ and $(Y, \mathcal{N}, \nu)$ be σ-finite measure spaces, f an $(\mathcal{M} \otimes \mathcal{N})$-measurable function on $X \times Y$,* and $1 \le p \le \infty$. If $f(\cdot, y) \in L^p(\mu)$ for almost every y and the function $y \mapsto \|f(\cdot, y)\|_p$ is in $L^1(\nu)$, then $f(x, \cdot) \in L^1(\nu)$ for almost every x, the function $x \mapsto \int f(x, y) \, d\nu(y)$ is in $L^p(\mu)$, and

$$(5.11) \qquad \left\| \int f(\cdot, y) \, d\nu(y) \right\|_p \le \int \|f(\cdot, y)\|_p \, d\nu(y).$$

It is enough to assume that $f \ge 0$ (by considering $|f|$ in place of f). When $p = 1$, the result is immediate from the Fubini-Tonelli theorem. When $1 < p < \infty$, the pth powers and pth roots get in the way of applying this theorem directly; instead, we integrate $\int f(x, y) \, d\nu(y)$ against a function $g \in L^q(\mu)$ where q is conjugate to p and then apply the Fubini-Tonelli theorem and Hölder's inequality:

$$
\begin{aligned}
(5.12) \qquad & \int \left[\int f(x, y) \, d\nu(y) \right] |g(x)| \, d\mu(x) \\
&= \iint f(x, y) |g(x)| \, d\mu(x) \, d\nu(y) \\
&\le \|g\|_q \int \left[\int f(x, y)^p \, d\mu(x) \right]^{1/p} d\nu(y) \\
&= \|g\|_q \int \|f(\cdot, y)\|_p \, d\nu(y).
\end{aligned}
$$

By Theorem 5.9, the quantity on the left of (5.11) is the supremum of the quantity on the left of (5.12) over all $g \in L^q(\mu)$ with $\|g\|_q \le 1$, so the result follows. (The case $p = \infty$ is almost a triviality.)

5.2   SPACES OF CONTINUOUS FUNCTIONS

Let X be an LCH space. We denote by $C(X)$ the space of all continuous complex-valued functions on X and by $BC(X)$ the space of all bounded functions in $C(X)$, and and we recall from §3.5 that $C_c(X)$ is the space of all functions in $C(X)$ that vanish outside a compact set. Thus we have

$$C_c(X) \subset BC(X) \subset C(X),$$

with equality if X is compact. If we wish to consider real-valued functions only, we shall indicate this with a superscript $\mathbb{R}$: $C^{\mathbb{R}}(X)$, etc.

The canonical norm on $BC(X)$ is the *uniform norm* or *supremum norm*,

$$\|f\|_u = \sup_{x \in X} |f(x)|,$$

convergence with respect to which is uniform convergence on X. The familiar fact that the uniform limit of a sequence of continuous functions is continuous remains true in this general setting (with the same easy proof), so $BC(X)$ is a Banach space.

When X is not compact, $C_c(X)$ is not a closed subspace of $BC(X)$. Its closure is the space of continuous functions f that *vanish at infinity* in the sense that $\{x : |f(x)| \geq \epsilon\}$ is compact for every $\epsilon > 0$. We denote this space by $C_0(X)$:

$$C_0(X) = \{f \in C(X) : \{x : |f(x)| \geq \epsilon\} \text{ is compact for all } \epsilon > 0\}.$$

It is easy to check that if $\{f_n\} \subset C_c(X)$ and $f_n \to f$ uniformly, then $f \in C_0(X)$. Conversely, if $f \in C_0(X)$, one uses Urysohn's lemma to find functions $g_n \in C_c(X)$ with $0 \leq g_n \leq 1$ and $g_n = 1$ on the set where $|f| \geq 1/n$; then $fg_n \to f$ uniformly. The phrase "vanish at infinity" is justified by another characterization of $C_0(X)$: it is the space of all continuous functions f on X that can be extended continuously to the one-point compactification of X (see §1.4) by setting $f(\infty) = 0$.

If X is not compact, the most generally useful topology on $C(X)$ is the topology of uniform convergence on compact sets, introduced in §4.3, and this space is most tractable when X is a σ-compact LCH space. Under this condition, there is a sequence $\{K_n\}$ of compact sets such that

(5.13) $X = \bigcup K_n$ and $K_n \subset \text{interior}(K_{n+1})$ for all n;

the topology of uniform convergence on compact sets is then defined by the countable family of seminorms $p_n(f) = \sup_{x \in K_n} |f(x)|$. This topology is therefore first countable, so it suffices to consider sequential convergence. Moreover, $C(X)$ is complete in the sense that if the sequence $\{f_n\}$ is uniformly Cauchy on each K_n, its limit again belongs to $C(X)$.

The main results of this section are three big theorems: a compactness theorem, an approximation theorem, and a duality theorem.

The compactness theorem is, along with Tychonoff's theorem, one of the few tools for establishing compactness in infinite-dimensional spaces. To state it, we need some terminology. Let $\mathcal{F}$ be a subset of $C(X)$. $\mathcal{F}$ is called *pointwise bounded* if $\{f(x) : f \in \mathcal{F}\}$ is a bounded subset of $\mathbb{C}$ for each $x \in X$. $\mathcal{F}$ is called *equicontinuous* if for every $x \in X$ and $\epsilon > 0$ there is a neighborhood U of x such that $|f(y) - f(x)| < \epsilon$ for all $y \in U$ and all $f \in \mathcal{F}$. (When X is an open subset of Euclidean space $\mathbb{R}^n$ and $\mathcal{F}$ consists of continuously differentiable functions, the mean value theorem of calculus yields a useful sufficient condition for equicontinuity: for each $x \in X$ there should be a ball $B \subset X$ centered at x and a constant $C > 0$ such that $|\nabla f| \leq C$ on B for all $f \in \mathcal{F}$.)

5.14 The Arzelà-Ascoli theorem. *Let X be a compact Hausdorff space. If $\mathcal{F}$ is a pointwise bounded, equicontinuous subset of $C(X)$, then $\mathcal{F}$ is totally bounded with respect to the uniform norm.*

5.15 Corollary. *With X and $\mathcal{F}$ as above, the closure of $\mathcal{F}$ is compact in $C(X)$.*

5.16 Corollary. *Suppose X is a σ-compact LCH space. Every pointwise bounded, equicontinuous sequence in $C(X)$ has a subsequence that converges uniformly on compact sets.*

The idea of the proof of the Arzelà-Ascoli theorem is as follows: Given $\epsilon > 0$, one uses equicontinuity and the compactness of X to show that there is a finite set $\{x_1, \ldots, x_m\} \subset X$ such that the value of every $f \in \mathcal{F}$ at an arbitrary $x \in X$ is within ϵ of its value at some x_j, and then uses pointwise boundedness to show that there is a finite set $\{y_1, \ldots, y_n\} \subset \mathbb{C}$ such that the value of every $f \in \mathcal{F}$ at each x_j is within ϵ of some y_j. This easily yields a covering of $\mathcal{F}$ by finitely many sets of diameter at most 4ϵ. Corollary 5.15 follows immediately. To establish Corollary 5.16, let $\{K_n\}$ be as in (5.13). If $\{f_n\}$ is pointwise bounded and equicontinuous, by the Arzelà-Ascoli theorem there is a subsequence that converges uniformly on K_1, a sub-subsequence that converges uniformly on K_2, and so forth; a diagonal process then yields a subsequence that converges uniformly on every K_n and hence on every compact set.

The approximation theorem is a vast generalization of the classic theorem of Weierstrass that every continuous function on a compact interval $[a, b] \subset \mathbb{R}$ is the uniform limit of polynomials. It depends on the fact that $C(X)$ is not merely a vector space but an *algebra*; that is, it is closed not only under addition and multiplication by scalars but also under multiplication of two functions, $(f, g) \mapsto fg$. Its proof (a beautiful but intricate argument that we omit) also depends strongly on the order structure of the real numbers, so it is most natural to state it in terms of real-valued functions; we shall state the complex version as a corollary. One more bit of terminology: a subset $\mathcal{F}$ of $C(X)$ *separates points* if for every pair of distinct points $x, y \in X$ there is an $f \in \mathcal{F}$ such that $f(x) \neq f(y)$.

5.17 The Stone-Weierstrass theorem. *Suppose X is a compact Hausdorff space and $\mathcal{A}$ is a subalgebra of $C^{\mathbb{R}}(X)$ that separates points. If there is* *an $x_0 \in X$ such that $f(x_0) = 0$ for all $f \in \mathcal{A}$, then $\mathcal{A}$ is dense in $\{f \in C^{\mathbb{R}}(X) : f(x_0) = 0\}$. Otherwise, $\mathcal{A}$ is dense in $C^{\mathbb{R}}(X)$.*

This theorem is false as it stands if $C^{\mathbb{R}}(X)$ is replaced by $C(X)$. For example, if X is the closed unit disc in $\{z : |z| \le 1\} \subset \mathbb{C}$ and $\mathcal{A}$ is the algebra of polynomials in one complex variable z (considered as functions on X), the closure of $\mathcal{A}$ in $C(X)$ is not all of $C(X)$ but the subalgebra of functions that are holomorphic (complex-analytic) on the open disc $\{z : |z| < 1\}$. What is needed to obtain a valid result is the assumption that $\mathcal{A}$ is closed under taking real and imaginary parts, or equivalently under complex conjugation, so that one can reduce to the real case.

5.18 Corollary. *Suppose X is a compact Hausdorff space and $\mathcal{A}$ is a sub-algebra of $C(X)$ that separates points and is closed under complex conjugation. If there is an $x_0 \in X$ such that $f(x_0) = 0$ for all $f \in \mathcal{A}$, then $\mathcal{A}$ is dense in $\{f \in C(X) : f(x_0) = 0\}$. Otherwise, $\mathcal{A}$ is dense in $C(X)$.*

Finally, the duality theorem. Let X be a compact Hausdorff space, and let $M(X)$ be the set of regular complex Borel measures on X. (A complex measure is called *regular* if the positive and negative parts of its real and imaginary parts are regular.) It is easily seen that $M(X)$ a vector space and the functional

$$\|\mu\| = |\mu|(X)$$

is a norm on it. (Recall that $|\mu|$ is the total variation of μ; see §2.5.)

Every $\mu \in M(X)$ defines a bounded linear functional ϕ_μ on $C(X)$ by

$$\phi_\mu(f) = \int f \, d\mu,$$

and we have $\|\phi_\mu\| \le \|\mu\|$ because

$$\left| \int f \, d\mu \right| \le \int |f| \, d|\mu| \le \|f\|_u \|\mu\|.$$

In fact, $\|\phi_\mu\| = \|\mu\|$, because the last inequality is an equality if $f = \overline{d\mu/d|\mu|}$. (The latter function might not be continuous, but it can be approximated by continuous functions by Lusin's theorem.) Thus the map $\mu \mapsto \phi_\mu$ is a norm-preserving injection of $M(X)$ into $C(X)^*$.

On the other hand, the Riesz representation theorem tells us that every positive $\phi \in C(X)^*$ (in the sense that $\phi(f) \ge 0$ whenever $f \ge 0$) is of the form ϕ_μ for some positive $\mu \in M(X)$. The full picture emerges from the fact that if $\phi \in C(X)^*$ is real (meaning that $\phi(f)$ is real whenever f is real) then ϕ has a "Jordan decomposition" as the difference of two positive linear functionals. Every $\phi \in C(X)^*$ is of the form $\phi_1 + i\phi_2$ where ϕ_1, ϕ_2 are real, so we are led to the following result:

5.19 Theorem. *If X is a compact Hausdorff space, the map $\mu \mapsto \phi_\mu$ is a norm-preserving isomorphism from $M(X)$ to $C(X)^*$.*

This theorem easily yields a similar characterization of $C_0(X)^*$ when X is an LCH space. Let $\widehat{X}$ be the one-point compactification of X. We have $C(\widehat{X}) \cong C_0(X) \oplus C$ where C is the one-dimensional space of constant functions, and $C(\widehat{X})^* = M(\widehat{X})$. It follows easily that $C_0(X)^*$ can be identified with the set of all $\mu \in M(\widehat{X})$ such that $\mu(\{\infty\}) = 0$, which in turn is just $M(X)$, the space of regular complex Borel measures on X. (To avoid confusion, recall that a positive measure qualifies as a complex measure only if it is finite. Every positive measure, finite or not, defines a linear functional on $C_c(X)$, but only the finite ones give finite integrals for every function in $C_0(X)$.) In short:

5.20 Corollary. *If X is an LCH space, the map $\mu \mapsto \phi_\mu$ is a norm-preserving isomorphism from $M(X)$ to $C_0(X)^*$.*

CHAPTER 6

TOPICS IN ANALYSIS ON EUCLIDEAN SPACE

In this chapter we present a few basic applications of the abstract ideas and results from the preceding chapters in the concrete setting of the analysis of functions of one or several real variables. There is much more to be said; we are merely scratching the surface of a vast subject that has undergone a vigorous development in the last century. Some references for more extensive treatments include Dym and McKean [2], Strichartz [21], and Stein [20].

We begin with a few matters of notation. First, we shall denote the integral of a function f on $\mathbb{R}^n$ with respect to Lebesgue measure by $\int f$ or $\int f(x)\,dx$ rather than $\int f\,d\lambda$, and we denote the L^p spaces with respect to Lebesgue measure simply by L^p. Second, we define the translation operator τ_a for $a \in \mathbb{R}^n$ by

$$\tau_a f(x) = f(x - a).$$

(This τ_a differs from the τ_a in Theorem 3.6: that one acts on points in $\mathbb{R}^n$, whereas this one acts on functions on $\mathbb{R}^n$. Although one has a $+$ sign and the other has a $-$ sign, they have analogous geometric effects. For example, if $n = 1$ and $a > 0$, the τ_a in Theorem 3.6 shifts points to the right by a; the τ_a here shifts graphs of functions to the right by a.) Third, we introduce the multi-index notation for polynomials and partial derivatives. A *multi-index* is an n-tuple $\alpha = (\alpha_1, \ldots, \alpha_n)$ of nonnegative integers. If α is a multi-index and $x \in \mathbb{R}^n$, we define

$$x^\alpha = \prod_1^n x_j^{\alpha_j}, \qquad \partial^\alpha = \prod_1^n \left[\frac{\partial}{\partial x_j}\right]^{\alpha_j},$$

$$|\alpha| = \alpha_1 + \alpha_2 + \cdots + \alpha_n, \qquad \alpha! = \alpha_1!\alpha_2!\cdots\alpha_n!.$$

These conventions are all illustrated in the formula for the Taylor polynomial of order k of a smooth function f about a point x_0:

$$P_{k,f,x_0}(x) = \sum_{|\alpha| \leq k} \partial^\alpha f(x_0) \frac{(x - x_0)^\alpha}{\alpha!}.$$

If U is an open set in $\mathbb{R}^n$, we denote by $C^k(U)$ the space of all functions f on U such that $\partial^\alpha f$ exists and is continuous on U for all α with $|\alpha| \leq k$ (in which case the order of the differentiations in ∂^α is immaterial); we also say that such a function is *of class* C^k on U. We also define

$$C^\infty(U) = \bigcap_{k=0}^{\infty} C^k(U),$$

$$C_c^\infty(U) = \{f \in C^\infty(U) : \text{supp}(f) \text{ is a compact subset of } U\}.$$

When $U = \mathbb{R}^n$ we shall generally omit mentioning it; thus, $C^k = C^k(\mathbb{R}^n)$, etc.

6.1 CONVOLUTIONS

The *convolution* of two measurable functions f and g on $\mathbb{R}^n$ is the function $f * g$ defined by

$$(6.1) \qquad f * g(x) = \int f(x - y)g(y)\, dy = \int f(z)g(x - z)\, dz,$$

provided that the integrals converge. (The two integrals are equal by the obvious change of variable.) There are several different conditions on f and g that guarantee the well-definedness of $f * g$; the most important are the following.

- If f is locally integrable and g is bounded and has compact support, then $f * g(x)$ is defined for every x, and $f * g$ is a locally bounded function.

- If $f \in L^p$ and $g \in L^q$ where p and q are conjugate, then $f * g(x)$ exists for every x, and $f * g$ is a bounded function with $\|f * g\|_u \leq \|f\|_p \|g\|_q$. This follows immediately from Hölder's inequality and the invariance of Lebesgue measure under translations and reflections.

- If $f \in L^p$ and $g \in L^1$, then $f * g(x)$ exists for almost every x, $f * g \in L^p$, and $\|f * g\|_p \leq \|f\|_p \|g\|_1$ (*Young's inequality*). This follows from Minkowski's inequality for integrals:

$$\|f*g\|_p \le \int \|f(\cdot-y)\|_p |g(y)|\,dy = \|f\|_p \int |g(y)|\,dy = \|f\|_p \|g\|_1.$$

The equality of the two integrals in (6.1) shows that convolution is commutative: $f*g = g*f$. It is also associative ($f*(g*h) = (f*g)*h$), as one sees by an application of the Fubini-Tonelli theorem. In particular, L^1 is a commutative algebra under convolution. (This is the analysts' version, for the additive group $\mathbb{R}^n$, of the algebraists' "group algebra" of a finite or discrete group.)

Observe that $f*g = \int (\tau_y f) g(y)\,dy$ is a continuous linear combination of translates of f, with coefficients given by the values of g. If g is supported in a small ball centered at the origin, then the integral defining $f*g(x)$ involves only the values of f in a small ball centered at x, so that if f is continuous at x, $f*g(x)$ should be nearly $f(x)\int g(y)\,dy$. This is the idea behind the following important approximation theorem.

6.2 Theorem. *Suppose that* $\phi \in L^1$ *and* $\int \phi(x)\,dx = 1$, *and for* $t > 0$ *let* $\phi_t(x) = t^{-n}\phi(t^{-1}x)$.
a. *If* $f \in L^p$ ($1 \le p < \infty$), *then* $\|f*\phi_t - f\|_p \to 0$ *as* $t \to 0$.
b. *If* f *is bounded and uniformly continuous, then* $\|f*\phi_t - f\|_u \to 0$ *as* $t \to 0$.
c. *Suppose also that* $|\phi(x)| \le C(1+|x|)^{-n-\epsilon}$ *for some* $C, \epsilon > 0$. *If* $f \in L^p$ ($1 \le p \le \infty$), *then* $f*\phi_t(x) \to f(x)$ *for every* x *in the Lebesgue set of* f — *in particular, for almost every* x, *and for every* x *at which* f *is continuous.*

The behavior of ϕ_t as $t \to 0$ is perhaps best seen by drawing a sketch. Replacement of x by $t^{-1}x$ in the argument of ϕ causes ϕ_t to be concentrated closer and closer to the origin as $t \to 0$, and the factor t^{-n} compensates for this compression so that $\int \phi_t(x)\,dx = 1$ for all t.

The essence of the proof of parts (a) and (b) is the following calculation. Since $\int \phi_t(y)\,dy = 1$ for all t,

$$f*\phi_t(x) - f(x) = \int [f(x-y) - f(x)]\phi_t(y)\,dy$$

$$= \int [f(x-tz) - f(x)]\phi(z)\,dz$$

$$= \int [\tau_{tz}f(x) - f(x)]\phi_t(z)\,dz,$$

so by Minkowski's inequality for integrals,

$$\|f * \phi_t - f\|_p \le \int \|\tau_{tz} f - f\|_p |\phi(z)| \, dz,$$

and likewise with $\|\cdot\|_p$ replaced by $\|\cdot\|_u$. Parts (a) and (b) then follow from the dominated convergence theorem provided we know that $\|\tau_{tz} f - f\|_p \to 0$ or $\|\tau_{tz} f - f\|_u \to 0$ as $t \to 0$. For the uniform norm this is just the definition of uniform continuity. For the L^p norm, it follows from these three facts: (i) functions in $C_c(\mathbb{R}^n)$ are uniformly continuous; (ii) uniform convergence implies L^p convergence on bounded sets; (iii) $C_c(\mathbb{R}^n)$ is dense in L^p. The proof of part (c) is more difficult.

Another important property of convolution is that if either f or g has some differentiability, then so does $f * g$. If g is smooth, then

$$(6.3) \qquad \partial^\alpha (f * g)(x) = \partial^\alpha \int f(y) g(x - y) \, dy$$

$$= \int f(y)(\partial^\alpha g)(x - y) \, dy = f * (\partial^\alpha g)(x),$$

under any conditions that permit the differentiation under the integral sign — for example, if g is of class C^k and $\partial^\alpha g$ is bounded for $|\alpha| \le k$, and $f \in L^1$ (see Theorem 2.14). Likewise, if f is smooth, we have $\partial^\alpha (f * g) = (\partial^\alpha f) * g$. Combining this with Theorem 6.2, we obtain results about approximation of arbitrary L^p functions or continuous functions by smooth functions. In particular:

6.4 Proposition. C_c^∞ *is dense in L^p for $1 \le p < \infty$ (in the L^p norm), and also dense in $C_0(\mathbb{R}^n)$ (in the uniform norm).*

Indeed, every function in L^p or C_0 can be approximated in the appropriate norm by functions of compact support, and the latter can be approximated by functions in C_c^∞ by convolving them with functions in C_c^∞. The only thing needed to make this work is a single $\phi \in C_c^\infty$ with $\int \phi(x) \, dx = 1$ to use in Theorem 6.2; for this purpose we can use

$$\phi(x) = \begin{cases} c \exp[1/(|x|^2 - 1)] & \text{if } |x| < 1, \\ 0 & \text{otherwise}, \end{cases}$$

where c is chosen to make $\int \phi(x) \, dx = 1$.

We also have a smooth version of Urysohn's lemma:

6.5 Proposition. *Suppose $K \subset U \subset \mathbb{R}^n$ where K is compact and U is open. There exists $f \in C_c^\infty$ such that $0 \le f \le 1$, $f = 1$ on K, and $f = 0$ outside U.*

Let δ be the minimum distance from points in K to points in $\mathbb{R}^n \setminus U$ (which is positive since K is compact), let V be the set of points whose distance to K is less than $\delta/2$, and choose a nonnegative $\phi \in C_c^\infty$ such that $\int \phi(x)\,dx = 1$ and $\phi(x) = 0$ for $|x| \geq \delta/2$; then $f = \chi_V * \phi$ does the job.

6.2 FOURIER SERIES AND TRANSFORMS

Apart from Taylor series, the most important series expansions of functions are the Fourier series and their relatives. The basic Fourier series that we discuss here pertain to periodic functions on the real line — that is, functions f for which there exists a number $a \neq 0$ such that $\tau_a f = f$, in which case f is called *a-periodic*. It suffices to take $a = 1$, as the general case is then obtained by rescaling, and we proceed to do so. A 1-periodic function, then, is really a function on the quotient group

$$\mathbb{T} = \mathbb{R}/\mathbb{Z}.$$

As a topological space, $\mathbb{T}$ is homeomorphic to the unit circle in the complex plane via the map $t \mapsto e^{2\pi i t}$, and we take the measure on $\mathbb{T}$ to be arc length divided by 2π. Alternatively, as far as measure and integration (but not topology) go, we can identify $\mathbb{T}$ with any interval of unit length, say $[-\frac{1}{2}, \frac{1}{2})$, and the measure on $\mathbb{T}$ is Lebesgue measure on this interval. (In other words, every 1-periodic function is determined by its restriction to $[-\frac{1}{2}, \frac{1}{2})$, and every function on $[-\frac{1}{2}, \frac{1}{2})$ extends uniquely to a 1-periodic function; in this way we identify $L^p(\mathbb{T})$ with $L^p([-\frac{1}{2}, \frac{1}{2}))$.)

The fundamental fact is the following:

6.6 Theorem. *For $n \in \mathbb{Z}$, let $e_k(t) = e^{2\pi i k t}$. Then $\{e_k\}_{-\infty}^\infty$ is an orthonormal basis for $L^2(\mathbb{T})$.*

The verification of orthonormality is an easy exercise in calculus. Since $e_j e_k = e_{j+k}$, the finite linear combinations of the e_k's (called *trigonometric polynomials*) form an algebra, which separates points on $\mathbb{T}$ (e_1 already does this), is closed under complex conjugation (since $\bar{e}_k = e_{-k}$), and contains nonvanishing functions (all the e_k's). By the Stone-Weierstrass theorem, the trigonometric polynomials are uniformly dense in $C(\mathbb{T})$, and hence also dense in $L^2(\mathbb{T})$ in the L^2 norm. It follows that $\{e_k\}_{-\infty}^\infty$ is an orthonormal basis.

For every $f \in L^2(\mathbb{T})$, then, we have

$$(6.7) \quad f = \sum_{-\infty}^\infty \widehat{f}(k) e_k, \qquad \widehat{f}(k) = \langle f, e_k \rangle = \int_{-1/2}^{1/2} f(t) e^{-2\pi i k t}\,dt,$$

where the series — the *Fourier series* of f — converges in the L^2 norm. The numbers $\widehat{f}(k)$ are called the *Fourier coefficients* of f. The integral defining $\widehat{f}(k)$ makes sense if f is merely in $L^1(\mathbb{T})$ (which properly includes $L^2(\mathbb{T})$), and $|\widehat{f}(k)| \leq \|f\|_1$; however, the interpretation of the series $\sum \widehat{f}(k)e_k$ is more problematic in this case. (See Theorem 6.9 below.)

If f is differentiable, there is a simple relation between the Fourier coefficients of f and those of its derivatives. Suppose f is of class C^M. For $m \leq M$, an m-fold integration by parts yields

$$\widehat{f^{(m)}}(k) = \int_{-1/2}^{1/2} f^{(m)}(t)e^{-2\pi ikt}\,dt$$

$$= \int_{-1/2}^{1/2} f(t)(2\pi ik)^m e^{-2\pi ikt}\,dt = (2\pi ik)^m \widehat{f}(k).$$

(The boundary terms all vanish by periodicity. Here it is important that f should be of class C^M as a periodic function on $\mathbb{R}$, not just as a function on $[-\frac{1}{2}, \frac{1}{2})$: the values of f and its derivatives at $-\frac{1}{2}$ and $\frac{1}{2}$ must match up.)

In particular, if f is of class C^1, we have $\widehat{f}(k) = \widehat{f'}(k)/2\pi ik$ for $k \neq 0$, and hence, by the Cauchy-Schwarz inequality and the Parseval identity,

$$\sum_{k \neq 0} |\widehat{f}(k)| \leq \left[\sum_{k \neq 0} \frac{1}{(2\pi k)^2}\right]^{1/2} \left[\sum_{k \neq 0} |\widehat{f'}(k)|^2\right]^{1/2} = C\|f'\|_2 < \infty.$$

Therefore:

6.8 Proposition. *If f is periodic and of class C^1, the Fourier series of f converges to f absolutely and uniformly.*

The study of the convergence of Fourier series of rougher functions (apart from the L^2 theory) is more delicate. Here is a summary of some of the most important results.

6.9 Theorem. *Let f be a periodic function on $\mathbb{R}$, and let*

$$S_n(t) = \sum_{-n}^{n} \widehat{f}(k)e^{2\pi ikt}$$

be the nth symmetric partial sum of its Fourier series.

a. If f is of bounded variation on $[0, 1]$, then

$$\lim_{n \to \infty} S_n(t) = \tfrac{1}{2}[f(t-) + f(t+)]$$

(defined as in (3.16)) for every t; in particular, $\lim_{n \to \infty} S_n(t) = f(t)$ for every t at which f is continuous.

b. There exists $f \in C(\mathbb{T})$ such that $\{S_n(0)\}$ diverges.

c. There exists $f \in L^1(\mathbb{T})$ such that $\{S_n(t)\}$ diverges for every t.

d. If $f \in L^p(\mathbb{T})$ with $1 < p < \infty$, then $\lim_{n \to \infty} \|S_n - f\|_p = 0$.

e. If $f \in L^p(\mathbb{T})$ with $p > 1$, then $\lim_{n \to \infty} S_n(t) = f(t)$ for almost every t.

All these results except (e) are in Zygmund's classic treatise [23, §§II.8, VII.6, VIII.1, and VIII.4]; the proof of (a) is also in [6, §8.5]. Part (e) is a deep and relatively recent result; a proof can be found in Fefferman [4].

Even more important than Fourier series is the Fourier transform, a way of expanding nonperiodic functions in terms of the exponentials $e^{2\pi i \omega t}$. To motivate it, we begin by observing that (6.7) can be adapted to deal with functions that are $2L$-periodic rather than 1-periodic by a change of variable, yielding

$$(6.10) \qquad f = \frac{1}{2L} \sum_{-\infty}^{\infty} c_k^L e^{\pi i k(\cdot)/L}, \qquad c_k^L = \int_{-L}^{L} f(t) e^{-\pi i k t/L} \, dt.$$

Now suppose that $f \in C_c(\mathbb{R})$. We take L large enough so that f vanishes outside $[-L, L]$ and apply (6.10) to the $2L$-periodic function that agrees with f on $[-L, L]$, obtaining a series expansion of f that is valid on the interval $[-L, L]$. Since f vanishes outside $[-L, L]$, the integral $\int_{-L}^{L}$ in (6.10) can be replaced by $\int_{-\infty}^{\infty}$, so on setting $\Delta\omega = 1/2L$ and $\omega_k = k/2L = k\Delta\omega$, we can rewrite (6.10) as

$$f = \sum_{-\infty}^{\infty} \widehat{f}(\omega_k) e^{2\pi i \omega_k (\cdot)} \Delta\omega \text{ on } [-L, L], \qquad \widehat{f}(\omega_k) = \int_{-\infty}^{\infty} f(t) e^{-2\pi i \omega_k t} \, dt.$$

This series looks very much like a Riemann sum: it suggests that in the limit as $L \to \infty$ we should obtain

$$(6.11) \quad f(t) = \int_{-\infty}^{\infty} \widehat{f}(\omega) e^{2\pi i \omega t} \, d\omega, \qquad \widehat{f}(\omega) = \int_{-\infty}^{\infty} f(t) e^{-2\pi i \omega t} \, dt.$$

In fact, this is correct when suitably interpreted.

We proceed to make things precise, and at the same time we generalize to functions of several real variables. If $f \in L^1(\mathbb{R}^n)$, the *Fourier transform* of f is the function $\widehat{f}$ defined by

$$\widehat{f}(\xi) = \int f(x) e^{-2\pi i \xi \cdot x} \, dx.$$

(Note: Many people replace $e^{-2\pi i \xi \cdot x}$ by $e^{-i \xi \cdot x}$ in the formula defining $\widehat{f}$. The factor of 2π must then reappear in other places in the formulas relating to $\widehat{f}$, and different people use different conventions in this respect.) The function $\widehat{f}$ is clearly bounded, with $\|\widehat{f}\|_u \le \|f\|_1$, and it is continuous by Theorem 2.14. In fact, $\widehat{f} \in C_0(\mathbb{R}^n)$ (the *Riemann-Lebesgue lemma*); one verifies this first when f is smooth and compactly supported by an integration-by-parts argument, and then uses the density of such f's in L^1.

The calculation of specific Fourier transforms is something of an art. Here we just mention one particularly important example:

6.12 Proposition. *The function* $\phi(x) = e^{-\pi|x|^2}$ *is its own Fourier transform.*

There are several ways to prove this; here's the most elementary one. Since $\phi(x) = \prod_1^n e^{-\pi x_j^2}$ and $e^{2\pi i \xi \cdot x} = \prod_1^n e^{-2\pi i \xi_j x_j}$, the integral defining $\widehat{\phi}(\xi)$ is a product of one-dimensional integrals, so it suffices to assume $n = 1$. In that case, by differentiating under the integral and integrating by parts, one can check that $d\widehat{\phi}(\xi)/d\xi = -2\pi\xi\widehat{\phi}(\xi)$, whence $\widehat{\phi}(\xi) = ce^{-\pi\xi^2}$ for some constant c; and $c = 1$ since $\widehat{\phi}(0) = \int e^{-\pi x^2}\, dx = 1$.

Now, (6.11) suggests that we should also consider the *adjoint Fourier transform* $g \mapsto g^\vee$ defined for $g \in L^1$ by

$$g^\vee(x) = \int g(\xi)e^{2\pi i \xi \cdot x}\, d\xi = \widehat{g}(-x),$$

and that in fact we should have $f = (\widehat{f})^\vee$. The trouble is that $\widehat{f}$ is usually not in L^1, so the integral defining $(\widehat{f})^\vee$ may not converge. (A simple example: $f = \chi_{[-1,1]}$, for which $\widehat{f}(\xi) = (\sin 2\pi\xi)/\pi\xi$.) The remedy for this is to introduce an extra factor into the formula for $(\widehat{f})^\vee$ to force convergence, and then remove it by a suitable limiting process.

6.13 The Fourier inversion theorem. *Let* Φ *be a function such that* $\Phi \in L^1 \cap C_0$, $\Phi(0) = 1$, *and* $\Phi^\vee \in L^1$. *Given* $f \in L^1$, *for* $t > 0$ *let*

$$f^t(x) = \int \widehat{f}(\xi)\Phi(t\xi)e^{2\pi i \xi \cdot x}\, d\xi.$$

Then

a. $\|f^t - f\|_1 \to 0$ *as* $t \to 0$, *and if also* $f \in L^p$ ($1 < p < \infty$), *then* $\|f^t - f\|_p \to 0$ *as* $t \to 0$.

b. *If* f *is bounded and uniformly continuous, then so is* f^t, *and* $f^t \to f$ *uniformly as* $t \to 0$.

c. *If* $\widehat{f} \in L^1$, *then* f *agrees almost everywhere with a continuous function* $\widetilde{f}$, *and* $(\widehat{f})^\vee = \widetilde{f}$.

An example of a Φ satisfying the given conditions is $\Phi(\xi) = e^{-\pi|\xi|^2}$, by Proposition 6.12. To prove (a) and (b), one performs some calculations to show that $f^t = f * \phi_t$, where $\phi = \Phi^\vee$ and ϕ_t is defined as in Theorem 6.2; the results then follow from that theorem. Part (c) follows by letting $t \to 0$ and applying the dominated convergence theorem.

There is an analogue of parts (a) and (b) of Theorem 6.13 for Fourier series that yields the classical summability methods for Fourier series; see [6, §8.4]. On the other hand, the analogue of Theorem 6.6 for Fourier transforms is the following:

6.14 The Plancherel theorem. *If $f, g \in L^1 \cap L^2$, we have*

$$\int f(x)\overline{g(x)}\, dx = \int \widehat{f}(\xi)\overline{\widehat{g}(\xi)}\, d\xi.$$

Consequently, the transforms $f \mapsto \widehat{f}$ and $f \mapsto f^\vee$ extend to mutually inverse unitary operators on L^2.

The essence of the proof, apart from a few technicalities about convergence, is as follows. For any $f, h \in L^1$ we have

$$(6.15) \qquad \int f\widehat{h} = \iint f(x)h(\xi)e^{-2\pi i \xi \cdot x}\, d\xi\, dx = \int \widehat{f}h.$$

Let $h = \overline{\widehat{g}}$; then by the inversion theorem,

$$\widehat{h}(x) = \int e^{-2\pi i \xi \cdot x}\overline{\widehat{g}(\xi)}\, d\xi = \int \overline{e^{2\pi i \xi \cdot x}\widehat{g}(\xi)}\, d\xi = \overline{g(x)},$$

so that $\int f\overline{g} = \int f\widehat{h} = \int \widehat{f}h = \int \widehat{f}\overline{\widehat{g}}$.

Much of the usefulness of the Fourier transform comes from its interaction with translations, linear transformations, differentiations, and convolution. Here is a summary of the results.

6.16 Proposition. *Suppose $f, g \in L^1(\mathbb{R}^n)$.*

a. *$(\tau_y f)\widehat{}(\xi) = e^{-2\pi i \xi \cdot y}\widehat{f}(\xi)$ and $\tau_\eta(\widehat{f}) = \widehat{h}$ where $h(x) = e^{2\pi i \eta \cdot x}f(x)$.*

b. *If T is an invertible linear transformation of $\mathbb{R}^n$, then $(f \circ T)\widehat{} = |\det T|^{-1}\widehat{f} \circ (T^*)^{-1}$. In particular, if T is a rotation, then $(f \circ T)\widehat{} = \widehat{f} \circ T$.*

c. *If f is of class C^k, $\partial^\alpha f \in L^1$ for $|\alpha| \le k$, and $\partial^\alpha f \in C_0$ for $|\alpha| \le k-1$, then $(\partial^\alpha f)\widehat{}(\xi) = (2\pi i \xi)^\alpha \widehat{f}(\xi)$ for $|\alpha| \le k$.*

d. *If $x^\alpha f$ (i.e., the function $x \mapsto x^\alpha f(x)$) is in L^1 for $|\alpha| \le k$, then $\widehat{f}$ is of class C^k, and $\partial^\alpha \widehat{f} = [(-2\pi i x)^\alpha f]\widehat{}$.*

e. $(f * g)\widehat{} = \widehat{f}\,\widehat{g}$.

Parts (a) and (b) are proved by a change of variable, (c) by integration by parts, (d) by differentiation under the integral, and (e) by Fubini's theorem.

The fact that the Fourier transform converts differentiation into a simple algebraic operation makes it a powerful tool in the theory of differential equations. For a simple illustration of its utility, let us consider the initial value problem for the *heat equation*,

$$\frac{\partial u}{\partial t} = \sum_{1}^{n} \frac{\partial^2 u}{\partial x_j^2}, \qquad u(x,0) = f(x).$$

This is a model for heat flow in a homogeneous isotropic medium; $u(x,t)$ represents the temperature at position x and time t and $f(x)$ is the initial temperature. The Fourier transform with respect to the spatial variable x turns this into the elementary ordinary differential equation

$$\frac{\partial \widehat{u}}{\partial t}(\xi,t) = -4\pi^2|\xi|^2\widehat{u}(\xi,t), \qquad \widehat{u}(\xi,0) = \widehat{f}(\xi),$$

whose solution is $\widehat{u}(\xi,t) = \widehat{f}(\xi)e^{-4\pi^2|\xi|^2 t}$. In view of Proposition 6.16e and a rescaling of Proposition 6.12, we obtain

$$u(x,t) = f * g_t(x), \qquad g_t(x) = (4\pi t)^{-n/2}e^{-|\xi|^2/4t}.$$

We have arrived at this formula without worrying about hypotheses to guarantee the validity of the calculations, but once we have it, we can check directly that it works. An elementary calculation shows that $g(x,t) = g_t(x)$ satisfies the heat equation, and hence so does u, by (6.3). Moreover, an application of Theorem 6.2, with $\phi_t(x) = g_{t^2}(x)$, shows that $u(\cdot,t) \to f$ as $t \to 0$ (in various senses, depending on what hypotheses one assumes on f).

6.3   DISTRIBUTIONS

In many situations it is convenient to enlarge the universe of discourse to include certain "generalized functions" that may be more singular than ordinary functions. Engineers and physicists began doing so more than a century ago, but it took mathematicians a while to catch up by developing a rigorous theory. The key idea is to change the focus from the pointwise values $f(x)$ to the integrals $\int f\phi$ as ϕ ranges over a suitable family of "test functions" — that is, to consider f as a linear functional on some function space. We have already seen that L^p functions are determined by their action as linear functionals on the conjugate space L^q. More generally, if f is a locally integrable function, the map $\phi \mapsto \int f\phi$ is a linear functional on the space of bounded measurable functions that vanish outside a compact set, and we can recover the pointwise values $f(x)$, for almost every x, from the integrals $\int f\phi$ by Theorem 6.2. The idea behind generalized functions, then, is that one can also consider other linear functionals on such function spaces.

The simplest and most famous example of a generalized function is the "Dirac delta-function" δ, which is supposed to have the property that $\delta(x) = 0$ for all $x \neq 0$ but $\delta(0) = \infty$ in such a way that $\int \delta(x) \, dx = 1$. If one takes this description literally, δ is a mythical beast. However, what it really means is that if one integrates δ against any reasonable function ϕ, the result should be $\int \delta(x)\phi(x) \, dx = \phi(0) \int \delta(x) \, dx = \phi(0)$; and from this point of view one immediately sees that δ exists not as a function but as a measure, namely, the point mass at 0. (What fails to exist is the Radon-Nikodym derivative of this measure with respect to Lebesgue measure.)

The largest generally useful class of generalized functions is the space of *distributions* invented by Laurent Schwartz. To define it we take the space of "test functions" to be C_c^∞ equipped with the following notion of sequential convergence: If $\{\phi_k\}$ is a sequence in C_c^∞, we say that $\phi_k \to \phi$ in C_c^∞ if (i) $\partial^\alpha \phi_k \to \partial^\alpha \phi$ uniformly for all multi-indices α and (ii) the ϕ_k's (and ϕ) are all supported in a common compact set. (This notion of sequential convergence comes from a certain topology on C_c^∞, but the description of the latter is rather complicated and will not be needed here. See Rudin [18].)

A *distribution* (on $\mathbb{R}^n$) is a linear functional F on the space C_c^∞ of compactly supported smooth functions that is continuous in the sense that if $\phi_k \to \phi$ in C_c^∞, then $F(\phi_k) \to F(\phi)$. The space of all distributions is denoted (following Schwartz) by $\mathcal{D}'$. We equip $\mathcal{D}'$ with the weak* topology, so that a sequence $\{F_m\} \subset \mathcal{D}'$ converges to F in $\mathcal{D}'$ if and only if $F_m(\phi) \to F(\phi)$ for all $\phi \in C_c^\infty$. More generally, if U is an open set in $\mathbb{R}^n$, a distribution on U is a continuous linear functional on $C_c^\infty(U)$. For simplicity, we shall restrict attention to distributions on $\mathbb{R}^n$, but most of what follows can be generalized to distributions on other open sets.

Every locally integrable function f on $\mathbb{R}^n$ defines a distribution, namely $\phi \mapsto \int f\phi$, and every positive or complex regular Borel measure μ defines a distribution, namely $\phi \mapsto \int \phi \, d\mu$. (In both cases, the continuity of the linear functional follows from the dominated convergence theorem.) These

correspondences are one-to-one provided that we identify two locally integrable functions that are equal almost everywhere, and we shall still denote
the corresponding distributions by f and μ. Examples of distributions that
are not functions or measures are provided by linear functionals that involve
derivatives such as $\phi \mapsto \partial^\alpha \phi(0)$ ($|\alpha| > 0$).

To avoid confusion about arguments ($f(x)$ versus $f(\phi)$, or $\mu(E)$ versus $\mu(\phi)$), we shall henceforth denote the value of any distribution F on
a test function ϕ by $\langle F, \phi \rangle$ rather than $F(\phi)$. (This pairing between $\mathcal{D}'$
and C_c^∞ is linear in each variable, so there is a slight discrepancy with
our earlier notation for inner products.) We may sometimes write $\int F\phi$ or
even $\int F(x)\phi(x)\,dx$ for $\langle F, \phi \rangle$; the pretense that F is a genuine function
is sometimes notationally handy and causes no trouble if used with care.

Many useful operations can be extended from functions to distributions.
The general philosophy is this: suppose we have two linear operators T
and T' on C_c^∞ that are continuous in the sense that if $\phi_k \to \phi$ in C_c^∞
then $T\phi_k \to T\phi$ and $T'\phi_k \to T'\phi$ in C_c^∞, and that satisfy $\int (T\phi)\psi =
\int \phi(T'\psi)$ for all $\phi, \psi \in C_c^\infty$. If we regard ϕ and $T\phi$ as distributions, this
relation can be rewritten as $\langle T\phi, \psi \rangle = \langle \phi, T'\psi \rangle$. We can then extend T to
all distributions F by *defining* $\langle TF, \psi \rangle$ to be $\langle F, T'\psi \rangle$. (The linear functional TF thus defined is continuous by the continuity of T'.) Here is the
basic list of examples; in all of them, verification of continuity is an easy
exercise.

- Differentiation: If $T = \partial^\alpha$, then $T' = (-1)^{|\alpha|} \partial^\alpha$ (by integration by
  parts), so if $F \in \mathcal{D}'$, we define its derivative $\partial^\alpha F$ by $\langle \partial^\alpha F, \phi \rangle =
  (-1)^{|\alpha|} \langle F, \partial^\alpha \phi \rangle$.

- Multiplication by C^∞ functions: Given $g \in C^\infty$, let $T\phi = g\phi$. Then
  $T' = T$, so we define the product of a distribution F and a C^∞ function
  g by $\langle gF, \phi \rangle = \langle F, g\phi \rangle$.

- Translation: We have $(\tau_y)' = \tau_{-y}$, so we define the translate $\tau_y F$ of a
  distribution F by $\langle \tau_y F, \phi \rangle = \langle F, \tau_{-y}\phi \rangle$.

- Composition with linear maps: Given an invertible linear transformation
  S of $\mathbb{R}^n$, let $T\phi = \phi \circ S$. Then $T'\phi = |\det S|^{-1} \phi \circ S^{-1}$ (by Theorem
  3.6), so for $F \in \mathcal{D}'$ we define $F \circ S$ by $\langle F \circ S, \phi \rangle = |\det S|^{-1} \langle F, \phi \circ
  S^{-1} \rangle$.

- Convolution with test functions: Given $\phi, \psi \in C_c^\infty$, we have

$$\phi * \psi(x) = \int \phi(y)\psi(x - y)\,dy = \int \phi[\tau_x(\psi \circ R)],$$

where $Ry = -y$. We can therefore define the convolution $F * \psi$ for $F \in \mathcal{D}'$ as a continuous function on $\mathbb{R}^n$ by $F * \psi(x) = \langle F, \tau_x(\psi \circ R) \rangle$. In fact, $F * \psi$ is C^∞ and $\partial^\alpha(F * \psi) = F * (\partial^\alpha \psi)$. (For $|\alpha| = 1$, use the fact that the appropriate difference quotients of ψ converge to $\partial^\alpha \psi$ in C_c^∞, and then proceed by induction on $|\alpha|$.) It then follows easily from the definition of $\partial^\alpha F$ that we also have $\partial^\alpha(F * \psi) = (\partial^\alpha F) * \psi$ — just as for convolution of functions. The notion of convolution can be extended in various ways to situations where neither factor is a test function, such as the convolution of L^p functions discussed in §6.1.

Convolutions can be used to approximate distributions by smooth functions. Suppose $\phi \in C_c^\infty$ and $\int \phi = 1$, and let $\phi_t(x) = t^{-n}\phi(t^{-1}x)$ for $t > 0$. It is an immediate corollary of Theorem 6.2 that ϕ_t converges to the Dirac distribution δ (the point mass at 0) in $\mathcal{D}'$ as $t \to 0$; moreover, for any $F \in \mathcal{D}'$, $F * \phi_t \to F$ in $\mathcal{D}'$ as $t \to 0$.

The fact that distributions can be differentiated at will to yield other distributions is the source of much of their power and flexibility. We present a few examples to give the reader the flavor of the theory; in them, δ always denotes the point mass at the origin.

First, in dimension $n = 1$, let $f = \chi_{(0,\infty)}$. The distribution derivative f' is given by

$$\langle f', \phi \rangle = -\langle f, \phi' \rangle = -\int_0^\infty \phi'(x)\,dx = -\phi(x)\big|_0^\infty = \phi(0) = \langle \delta, \phi \rangle,$$

so $f' = \delta$. More generally, if f is a piecewise continuous function with a jump discontinuity at $x = a$, its distribution derivative will contain the term $[f(a+) - f(a-)]\tau_a\delta$.

As an illustration of the rules for manipulating distributions, still in dimension 1, we observe that $x\delta'(x) = -\delta(x)$. This statement is couched in the informal language where one pretends that distributions are functions; the precise meaning is that, with $\iota(x) = x$,

$$\langle \iota\delta', \phi \rangle = \langle \delta', \iota\phi \rangle = -\langle \delta, (\iota\phi)' \rangle$$
$$= -\frac{d}{dx}[x\phi(x)]_{x=0} = -\phi(0)$$
$$= -\langle \delta, \phi \rangle.$$

(If this relation seems a little mysterious, it may be instructive to consider smooth approximations. Think of δ as the limit as $\epsilon \to 0$ of a smooth function f_ϵ supported in $[-\epsilon, \epsilon]$ with a sharp spike of height $\approx 1/\epsilon$ in that interval. What does the graph of $f'_\epsilon(x)$ look like? What about $xf'_\epsilon(x)$?)

Our final example, in dimension $n = 3$, comes from mathematical physics. In units such that the proportionality constant in Coulomb's law is equal to 1, the electrostatic potential generated by a unit positive charge at the origin is $u(x) = |x|^{-1}$. On the other hand, the potential $v(x)$ generated by a distribution of charges is related to the density $\rho(x)$ of that distribution by Poisson's equation $\nabla^2 v(x) = -4\pi\rho(x)$, where $\nabla^2 = \sum_1^3 \partial^2/\partial x_j^2$ is the Laplace operator. For these two laws to be consistent, we should have $\nabla^2 u = -4\pi\delta$. To prove this, observe that u (which is locally integrable, by Proposition 3.9) is the limit in $\mathcal{D}'$ of the smooth functions $u^\epsilon(x) = (|x|^2 + \epsilon^2)^{-1/2}$. An elementary calculation shows that $\nabla^2 u^\epsilon(x) = \epsilon^{-3}\phi(x/\epsilon)$ where $\phi(x) = -3(|x|^2 + 1)^{-5/2}$, so Theorem 6.2 implies that $\nabla^2 u^\epsilon \to (\int \phi)\delta$. But Proposition 3.9 followed by the substitution $r = \tan\theta$ gives

$$\int \phi = -4\pi \int_0^\infty \frac{3r^2}{(r^2+1)^{5/2}}\, dr$$

$$= -4\pi \int_0^{\pi/2} 3\sin^2\theta\cos\theta\, d\theta$$

$$= -4\pi,$$

and we are done.

Our list of operations on distributions has one notable omission: the Fourier transform. The trouble is that the Fourier transform does not map C_c^∞ into itself, so it cannot be dualized to yield an operator on $\mathcal{D}'$. (In fact, if $\phi \in C_c^\infty$ is nonzero, $\widehat{\phi}$ cannot vanish on any nonempty open set. Suppose to the contrary that $\widehat{\phi} = 0$ on a neighborhood of ξ_0. By writing $\widehat{\phi}(\xi) = \int e^{-2\pi i(\xi-\xi_0)\cdot x} e^{-2\pi i\xi_0\cdot x}\phi(x)\, dx$, expanding the first exponential in its power series, and integrating term by term, one sees that $\widehat{\phi}(\xi)$ is the sum of its Taylor series about ξ_0 for all ξ; but that series vanishes identically.) The remedy is to consider a larger space of test functions, namely, the *Schwartz space* of "rapidly decreasing" functions,

$$\mathcal{S} = \{\phi \in C^\infty : x^\alpha \partial^\beta \phi \text{ is bounded for all multi-indices } \alpha, \beta\},$$

equipped with the topology defined by the family of norms $\|\phi\|_{\alpha,\beta} = \sup_x |x^\alpha \partial^\beta \phi(x)|$. The space $\mathcal{S}$ includes C_c^∞ as well as functions such as $x^\alpha e^{-\pi|x|^2}$ and $e^{-(|x|^2+1)^{1/10}}$.

A distribution is called *tempered* if it extends to a continuous linear functional on $\mathcal{S}$. (The extension is unique, because C_c^∞ is dense in $\mathcal{S}$ in the topology of $\mathcal{S}$.) Locally integrable functions are tempered if they grow at most polynomially at infinity, but functions such as $e^{|x|}$ are not. The operations on distributions that we discussed earlier can also be considered as operations on tempered distributions, except that the product of a tempered distribution and a C^∞ function f may not be tempered unless f is "slowly increasing" in the sense that $|\partial^\alpha f(x)| \leq C_\alpha (1 + |x|)^{N_\alpha}$ for all α.

It is an easy consequence of Proposition 6.16(c,d) that the Fourier transform does map $\mathcal{S}$ into itself, and in fact is an isomorphism of $\mathcal{S}$. We have $\int \widehat{\phi}\psi = \int \phi\widehat{\psi}$ for $\phi, \psi \in \mathcal{S}$ by (6.15), so we may extend the Fourier transform to all tempered distributions by the prescription $\langle \widehat{F}, \phi \rangle = \langle F, \widehat{\phi} \rangle$. The adjoint Fourier transform extends in the same way, $\langle F^\vee, \phi \rangle = \langle F, \phi^\vee \rangle$, and these two operators are inverses of each other on $\mathcal{S}'$ because they are so on $\mathcal{S}$. Moreover, the operational properties of the Fourier transform listed in Proposition 6.16 continue to hold on $\mathcal{S}'$.

We conclude with are some examples. First, the Fourier transform of the point mass δ is given by

$$\langle \widehat{\delta}, \phi \rangle = \langle \delta, \widehat{\phi} \rangle = \widehat{\phi}(0) = \int \phi(x)\,dx = \langle 1, \phi \rangle,$$

that is, $\widehat{\delta}$ is the constant function 1. It then follows from the extended version of Proposition 6.16(c) that $(\partial^\alpha \delta)\widehat{}$ is the monomial $(2\pi i \xi)^\alpha$, and then from the Fourier inversion formula that the Fourier transform of x^α is $(-2\pi i)^{-|\alpha|}\partial^\alpha \delta$. Moreover, we showed earlier that if $u(x) = |x|^{-1}$ on $\mathbb{R}^3$ then $\nabla^2 u = -4\pi\delta$. Applying the Fourier transform to this equality and using Proposition 6.16(c) again, we obtain $-4\pi^2|\xi|^2\widehat{u} = -4\pi$, so that $\widehat{u}$ is the locally integrable function $\widehat{u}(\xi) = 1/\pi|\xi|^2$.

Bibliography

[1] R. G. Bartle, Return to the Riemann integral, *Amer. Math. Monthly* **103** (1996), 625–632.

[2] H. Dym and H. P. McKean, *Fourier Series and Integrals*, Academic Press, New York, 1972.

[3] K. J. Falconer, *The Geometry of Fractal Sets*, Cambridge University Press, Cambridge, 1985.

[4] C. L. Fefferman, Pointwise convergence of Fourier series, *Annals of Math.* **98** (1973), 551–571.

[5] W. Fleming, *Functions of Several Variables* (2nd ed.), Springer, New York, 1977.

[6] G. B. Folland, *Real Analysis* (2nd ed.), John Wiley, New York, 1999.

[7] P. R. Halmos, *Naive Set Theory*, Van Nostrand, Princeton, NJ, 1960; reprinted by Springer, New York, 1974.

[8] J. Hennefeld, A nontopological proof of the uniform boundedness theorem, *Amer. Math. Monthly* **87** (1980), 217.

[9] F. Jones, *Lebesgue Integration on Euclidean Space*, Jones and Bartlett, Boston, 1993.

[10] T. W. Körner, *A Companion to Analysis*, American Mathematical Society, Providence, RI, 2004.

[11] S. G. Krantz, *A Guide to Real Variables*, Mathematical Association of America, Washington, DC, 2009.

[12] S. Lang, *Real and Functional Analysis* (3rd ed.), Springer, New York, 1993.

[13] R. M. McLeod, *The Generalized Riemann Integral*, Mathematical Association of America, Washington, DC, 1980.

[14] M. Reed and B. Simon, *Methods of Modern Mathematical Physics I: Functional Analysis*, Academic Press, New York, 1972.

[15] H. L. Royden, *Real Analysis* (3rd ed.), Macmillan, New York, 1988.

[16] W. Rudin, *Principles of Mathematical Analysis* (3rd ed.), McGraw-Hill, New York, 1976.

[17] W. Rudin, *Real and Complex Analysis* (3rd ed.), McGraw-Hill, New York, 1987.

[18] W. Rudin, *Functional Analysis* (2nd ed.), McGraw-Hill, New York, 1991.

[19] S. Saeki, A proof of the existence of infinite product probability measures, *Amer. Math. Monthly* **103** (1992), 682–683.

[20] E. M. Stein, *Singular Integrals and Differentiability Properties of Functions*, Princeton University Press, Princeton, NJ, 1970.

[21] R. S. Strichartz, *A Guide to Distribution Theory and Fourier Transforms*, CRC Press, Boca Raton, FL, 1994.

[22] K. Stromberg, The Banach-Tarski paradox, *Amer. Math. Monthly* **86** (1979), 151–161.

[23] A. Zygmund, *Trigonometric Series* (2 vols., reprinted in 1 vol.), Cambridge University Press, Cambridge, 1968.

INDEX

About the Author

Gerald B. Folland was born and raised in Salt Lake City, Utah. He received his bachelor's degree from Harvard University in 1968 and his doctorate from Princeton University in 1971. After two years at the Courant Institute, he moved to the University of Washington, where he is now professor of mathematics. Folland is the author of ten textbooks and research monographs in the areas of real analysis, harmonic analysis, partial differential equations, and mathematical physics.